CAMBRIDGE COUNTY GEOGRAPHIES

General Editor: F. H. H. GUILLEMARD, M.A., M.D.

T0352335

SUFFOLK

Cambridge County Geographies

SUFFOLK

by

W. A. DUTT

With Maps, Diagrams and Illustrations

Cambridge:
at the University Press
1909

CAMBRIDGE UNIVERSITY PRESS
Cambridge, New York, Melbourne, Madrid, Cape Town,
Singapore, São Paulo, Delhi, Mexico City

Cambridge University Press
The Edinburgh Building, Cambridge CB2 8RU, UK

Published in the United States of America by
Cambridge University Press, New York

www.cambridge.org
Information on this title: www.cambridge.org/9781107689909

© Cambridge University Press 1909

This publication is in copyright. Subject to statutory exception
and to the provisions of relevant collective licensing agreements,
no reproduction of any part may take place without the written
permission of Cambridge University Press.

First published 1909
First paperback edition 2012

*A catalogue record for this publication is available from the British
Library*

ISBN 978-1-107-68990-9 Paperback

This publication reproduces the text of the original edition of the
Cambridge County Geographies. The content of this publication has
not been updated. Cambridge University Press has no responsibility
for the accuracy of the geographical guidance or other information
contained in this publication, and does not guarantee that such
content is, or will remain, accurate.

CONTENTS

ILLUSTRATIONS

MAPS

The illustrations reproduced on pp. 5, 11, 17, 26, 35, 37, 45, 50, 60, 75, 81, 89, 90, 91, 92, 94, 95, 97, 102, and 112 are from photographs specially taken by Mr I. Dexter, of London; and those on pp. 13, 66, 68, 85, 104, and 106 are from photographs specially taken by Mr F. Manning, of Norwich. Mr H. Jenkins, of Lowestoft, supplied the views on pp. 9, 40, 47, 64, 71, and 129; and Mr G. S. Cousins, of Bury St Edmunds, those on pp. 79, 100 and 127. The author is indebted to Mr E. H. H. Hancox, of Nacton, for the view on p. 33; to Miss Prentice, Secretary of the Suffolk Sheep Society, for that on p. 62; and to Mr F. Woolnough, Curator of Ipswich Museum, for that of the Anglo-Saxon fibulae.

1. County and Shire. The Origin of Suffolk.

At the beginning of our study of the geography of Suffolk it will be as well to learn how the county came by its name. In order to do this we must look back almost to the beginning of its history, and then we shall see that there was a pre-existing geographical reason why the name of Suffolk, or some similar name, should have been given to it.

In the year 55 B.C., when Julius Caesar landed on the coast of Kent, he found different parts of England inhabited by different British tribes. For instance, the south-eastern portion, with which he first became acquainted, belonged to a people called the Cantii; the easternmost portion, which included Suffolk, was occupied by the Cenimagni or Iceni; while between the countries of these two tribes lay that of the Trinobantes.

We cannot now be quite sure of the precise boundaries of these tribal divisions of the country, but in the cases of some of them it is clear that they had natural boundaries in the shape of rivers, fens, or forests. The country of the Cantii was divided from that of the Regni to the west by a great forest covering the district now called the Weald; while the country of the Iceni was bounded on the north and on the east by the sea, on the north-west by the fens, and on the west by a large tract of forest.

The Romans were in possession of Britain till about 410 A.D. They then gradually withdrew and left the country practically defenceless, so that when the Saxons and Angles invaded it they had little difficulty in conquering and occupying it.

England was divided by the Saxons and Angles into several kingdoms. The boundaries of some of these new divisions differed from those of the countries which had been occupied by the British tribes, but the Saxon kingdom of Kent had nearly the same limits as the country of the Cantii, and the Anglian kingdom of East Anglia almost the same as the country of the Iceni.

If we look at a map of England, we notice at once that the country is divided into counties and shires. Of these the divisions with names ending in *shire* are portions or *shares* of the larger divisions which existed in Saxon times. For instance, Staffordshire was a part of the Saxon kingdom of Mercia, and Berkshire and Gloucestershire were parts of the kingdom of Wessex. As for the counties with names without the termination *shire*, most of these are old English kingdoms which have kept their original boundaries and in some cases their original names.

The kingdom of East Anglia was founded in 575 A.D. Its name implies that its inhabitants at that time were chiefly Angles. As we have said, we cannot be quite sure of its precise limits, but it was bounded on the west and north-west by the kingdom of Mercia and on the south by the kingdom of East Saxony, which included Essex. For many years it had its own kings, and it was often at war, not only with the Danes who invaded its coasts, but also with other Saxon kingdoms.

Now, if we look at the map again, we shall see that the kingdom of East Anglia was naturally divided into two portions by the Little Ouse and Waveney rivers. In course of time, its inhabitants began to describe themselves as North-folk or South-folk, the former being those who dwelt on the north and the latter those who dwelt on the south side of these two rivers.

So the county of Suffolk is really the country of the South-folk, while the county of Norfolk is the country of the North-folk of East Anglia.

2. General Characteristics. Position and Natural Conditions.

If we look at a map of England, we see at once that Suffolk, in consequence of the irregular projection of its north-eastern part, is the easternmost county of England. In our first chapter, we have learnt that the river Waveney is one of the boundaries between Suffolk and Norfolk. By tracing the course of the Waveney on the map we can understand why it is that Lowestoft Ness, the most easterly point of the kingdom, is in the former county instead of in the latter. For many miles of its course the

Waveney flows in an easterly direction, but when it has almost reached the coast at Lowestoft it turns northward, in which direction it flows for several miles before it enters the estuary of the Norfolk river Yare. In consequence of this bend in the river, a small portion of Suffolk, which looks as if it ought to belong to Norfolk, lies between the latter county and the sea. This portion was formerly known as the Island of Lothingland, and at the present time it is actually an island, though several bridges connect it with the rest of the county and with Norfolk.

It is evident therefore that Suffolk is a maritime county. It is bounded on the east by the North Sea, and it has about 50 miles of coast-line. In consequence of its exposed position it has always been liable to invasion by foreign enemies. The Romans built one, if not two, great fortresses to protect its coast, but when they left England the Saxons and Angles probably had no difficulty in landing on the shores of our county and taking possession of it. Subsequently the Danes or Northmen occasionally raided and plundered it. In the reign of Henry II, an army of Flemings landed on the coast, and marched into the heart of the county before being defeated by the King's troops. In 1667 the Dutch succeeded in landing about three thousand men near Felixstowe, but they were driven back to their ships. As recently as the beginning of the nineteenth century small forts called Martello towers were built at several points along the coast as a protection against invasion by the French.

The Suffolk port of Lowestoft is one of the principal British centres of the herring and trawl fisheries, but the maritime trade of the county is hardly so great as it used to be. Ipswich and Lowestoft are now its chief ports for carrying on import and export trade, but formerly there were three or four flourishing smaller ports along the coast. The trade of these latter towns has decayed or become quite extinct, owing to the destruction or silting-up of their harbours.

A Martello Tower

Suffolk was formerly noted for its weaving industries, which were introduced by the Flemings, who settled in considerable numbers in various parts of the county. The woollen industry is now so greatly decayed as to be almost extinct, and for many years Suffolk has been chiefly prominent as one of the principal agricultural counties of England. Over the greater part of its surface the soil is not very good, but skilful farming has made it very productive. The wheat grown on the stiff heavy lands is said to be the best in the kingdom, while the barley is also of excellent quality.

William Cobbett, the author of *Rural Rides*, who made a journey through Suffolk about a hundred years ago, commented on the "great number of farm-houses" and the neat state in which they were kept. "The land," he wrote, "is generally as clean as a garden ought to be; and though it varies a good deal as to lightness and stiffness, they make it all bear prodigious quantities of Swedish turnips; and on them pigs, sheep, and cattle all equally thrive. I did not observe a single poor miserable animal in the whole county." He adds that the cottages were also clean and comfortable, and that he did not see "one miserable hovel in which a labourer resided." This tribute to the skill and industry of the Suffolk farmers is as well deserved to-day as it was a century ago.

Suffolk is also noted as a stock-raising county and for its dairy-farming. Daniel Defoe, who visited the county in 1722, said that it was famous for producing the best butter in England. At the same time he remarked that it produced the worst cheese.

We have referred to the loss of trade Suffolk has sustained in consequence of sea encroachment having destroyed some of its small harbours and rendered others incapable of accommodating large ships. To compensate for this loss, the sea-side towns and villages have of late years become exceedingly popular with summer visitors, thousands of whom spend their holidays at Lowestoft, Felixstowe, Southwold, and other smaller watering-places.

In concluding this chapter, we may quote a passage from Robert Reyce, who wrote a *Breviary of Suffolk* in 1618. He says: "It is not amongst the least for which this shire is indebted to nature...that so quickly and commodiously it can vent and make return of such commodities which it affordeth; for if navigable rivers, diversity of commodious havens,...nearness unto the quickest and readiest markets of best trade, and with as little peril and small charge as any other shire, may be justly acknowledged the sole means of a profitable and commodious situation, then shall this shire, of all such as truly know it, justly deserve that true commendation."

With the exception of the reference to "the commodious havens," some of which, as we have said, no longer exist, this account of the advantageous situation of the county applies to it to-day.

3. Size. Shape. Boundaries.

Suffolk ranks twelfth in size among the counties of England. Its length, measured from Gorleston on its Norfolk border to Withersfield on the Cambridgeshire border, is about 70 miles; while its greatest breadth from east to west is 52 miles, and from north to south 32 miles. The entire area of the county is 952,710 acres, or 1,488 square miles. Of this area 7,299 acres or about $11\frac{1}{4}$ square miles is water. Now, the entire area of England, Wales and Scotland is 56,200,006 acres; so, as the land area of Suffolk is 945,411 acres, this county represents a 59th part of Great Britain.

In shape the county has been compared with an ancient galley or argosy, the rounded keel of which is represented by its southern border, while the high raised prow and stern of the ship are roughly outlined by the projecting portions of the county in the north-east and north-west respectively.

On the east and south-east Suffolk is bounded by the North Sea, the most northerly town on the coast being Gorleston, the most southerly Felixstowe. On the north it is bounded by Norfolk, from which it is separated by the rivers Little Ouse and Waveney. On the south it is bounded by Essex, from which it is separated by the river Stour. On the west it is bounded by Cambridgeshire, from which, in the north-western part of the county, it is separated by the river Lark.

It will thus be seen that nearly all the boundaries of Suffolk are natural ones. The only portions of the county where this is not the case are the part of its western border lying between Haverhill in the south-west and Worlington in the north-west, and a small part, less than two miles in extent, separating the head waters of the Waveney and Little Ouse.

Part of Newmarket, with the adjoining parish of Exning, forms an isolated portion of Suffolk, with which it is connected only by the main road from Thetford.

It is in consequence of the northern boundary of the county being a natural one that Suffolk claims, in the north-east, the small projecting portion, previously spoken of, known as the Island of Lothingland, which has in Lowestoft Ness the most easterly point of England.

The coast-line of Suffolk is remarkably unbroken and devoid of sharp promontories, the sole exception being Landguard Point, its southern extremity, which projects southward from Felixstowe and shelters Harwich Harbour. The only other prominent portions of the coast are Lowestoft Ness, Covehithe Ness, Thorpe Ness, and Orford Ness. Many years ago there was another point,

Lowestoft Harbour

between Covehithe and Southwold, called Easton Ness. This has now been washed away, but while it existed it was the most easterly point of the kingdom.

Nearly all, if not all, of the Suffolk nesses (an Anglo-Saxon form of the word noses) were far more prominent formerly than they are to-day, and as a result there were large sheltered bays between them. One of these bays was Sole Bay, off Southwold, where a great naval battle was fought in the seventeenth century. On most of the maps the only bay we now find marked is Hollesley Bay, south of Orford Ness. Here, also, English fleets used to ride safely at anchor in the days when our men-of-war were built of wood.

The two principal harbours on the Suffolk coast are Harwich Harbour, which lies between Suffolk and Essex, and Lowestoft Harbour. Only the smallest of coasting craft can enter the havens at Bawdsey, Orford, and Southwold.

4. Surface and General Features.

Robert Reyce, an old writer we have already quoted, thus quaintly describes the general aspect of the county of Suffolk. "This country, delighting in a continual evenness and plainness, is void of any great hills, high mountains, or steep rocks, notwithstanding the which it is not always so low, or flat, but that in every place it is severed and divided with little hills easy for ascent and pleasant rivers watering the low valleys, with a most beautiful prospect which ministreth unto the inhabitants a full choice of healthful and pleasant situations for their seemly houses."

Since this was written a good many persons have given us descriptions of Suffolk, and there seems to be a general agreement among them that the county, although boasting no very striking scenery, possesses many districts of alluring charm and quiet beauty. The great artist Constable, who was born in Suffolk, found in the neighbourhood of East Bergholt, his native village, the subjects of several of his masterpieces; while Gainsborough,

The Stour valley from Dedham church tower

another Suffolk painter, was similarly indebted to the scenery of his county.

The most picturesque parts of Suffolk are its river valleys. Five important rivers enter the sea along the coast, and of these the Stour, the Orwell, the Deben, and the Waveney are famous for their delightful scenery.

The Stour valley, in which Constable painted, yearly attracts many artists, while some of the upper reaches of the Waveney, which is one of the principal rivers of the Broads district, are very beautiful.

The late Canon Raven, an historian of Suffolk, says of its rivers that they have "a sleepy, home-like beauty of their own....The banks often fall sharp to the water edge, pleasantly bushed and flowered. Old locks and mills, as well as reed-beds and fords, have given words to the poet and drawn colour from the palette."

The central part of Suffolk consists of stiff heavy clay land. This is entirely an agricultural district, in which the greater part of the county's corn and green crops is raised. It includes what is generally called "High Suffolk," a vaguely defined district which may be said to consist of an elevated plain of clay land extending from Beccles in the north-east to Clare and Sudbury in the south-west. This district embraces a portion of the so-called Woodlands district; but nearly all the old woodlands of the county have now disappeared. In the parish of Sutton, near Woodbridge, there is a small tract of ancient oak forest known as Staverton Park.

Eastward of the clay lands an almost unbroken stretch of light soil extends along the coast from the river Orwell to the county boundary at Gorleston. Much of this light soil is uncultivated, and remains in the state of primitive heathland, but in many places it has been enriched by the addition of other soils and made productive. This tract of light soil used to be called the Sandy Lands.

In the north-west, too, there is a considerable tract of sandy land, forming part of what is called the Breck district, a name given to it because portions of it have been cultivated in large open fields called brecks. The sand and gravel of this district rest upon the chalk, an elevated ridge of which extends across the western part of the county, rising at Haverhill to a height of 352 feet

Flint-covered "breck," Brandon

above sea level. There are large heaths and warrens in the north-western part of Suffolk, which was formerly almost entirely open and barren, but its aspect has been altered of late years by the planting of long belts of firs.

A small portion of the great Fen district comes within the bounds of Suffolk where it borders the Breck lands. Most of it is in the parish of Mildenhall, which is the largest parish in the county, comprising nearly 17,000 acres of land. The Suffolk fens, like those of Cambridgeshire, are now drained and converted into arable and pasture land.

A portion of the popular Broads district also lies within the borders of this county. The river Waveney is navigable to Broadland cruising yachts as far as the town of Bungay.

Most of the marshlands of Suffolk border the rivers Waveney, Blyth, and Alde; but there are also tracts of marsh adjoining the coast at Benacre near Kessingland, and between Dunwich and Leiston. This last-named marsh district is called the Minsmere Level.

Along the coast, the land for the most part is low. From Gorleston to Lowestoft there is a line of sandy cliffs, and southward of Lowestoft cliffs of sand and clay, rising in places to a height of about 70 feet, extend as far as Kessingland. Further southward, at Covehithe, Southwold, and Dunwich there are low cliffs; but generally the land slopes gradually down to the shore. High ground, however, shelters the popular watering-places of Aldeburgh and Felixstowe. Where there are no cliffs, the beach is usually bordered by sandhills or banks of shingle.

5. Watershed. Rivers.

The watershed of Suffolk is an almost indistinguishable tract of slightly elevated land crossing the county in a south-westerly direction from the neighbourhood of Botesdale, near the sources of the Little Ouse and Waveney. As the land is fairly level, the rivers flow slowly until they become tidal and are influenced by the outrush of the tide. With the exception of the Little Ouse and the Lark, which flow towards the west and join the Norfolk and Cambridgeshire Ouse, all the main rivers take an easterly course towards the sea.

In order that we may learn something about the rivers, we will take each separately and trace its course from its source to the sea.

We will begin with those flowing in an easterly direction, taking the northernmost river first and ending with the southernmost.

The northernmost river is the Waveney, the boundary between Suffolk and Norfolk. It rises at South Lopham, near Diss in Norfolk, and flows by way of the towns of Bungay and Beccles to the neighbourhood of Lowestoft, where it turns northward and enters the estuary of the Norfolk river Yare, thus discharging its waters into the sea through Yarmouth harbour. Its only important tributary is the Dove, which joins it near Hoxne, but about three miles from Lowestoft it is connected with Oulton Broad by a navigable channel called Oulton Dyke. The Broad, which is about 130 acres in extent, is one of the best known of the sheets of water in the Broads district. It adjoins Lake Lothing, a tidal water connecting it with the harbour at Lowestoft. The Waveney is navigable to wherries and cruising yachts between Bungay and the junction of the river with the Yare.

Following the coast-line southward as far as Southwold, we come to the little river Blyth, which rises near Laxfield, about 8 miles west of Halesworth, and is joined by several small nameless tributaries before it reaches Southwold.

In our last chapter we referred to a tract of marshland lying south of Dunwich, called the Minsmere Level. These marshes border a small stream, the Minsmere, which rises near Ubbeston. On the bank of this stream is the village of Yoxford, which has been called the " Garden of Suffolk."

We now come to the Alde, a larger river having its source at Brundish. A little way below the village of Stratford St Andrew, it is joined by the Ore. It nearly reaches the sea at Aldeburgh, but it is there turned southward by a long narrow strip of sand and shingle. It then flows on past the little town of Orford, and after receiving the waters of the small Butley river it enters the sea at Hollesley Bay. Between Aldeburgh and the sea this river is sometimes called the Ore.

The next river, as we travel southward, is the charming Deben, which rises near Debenham in High Suffolk and flows by way of Wickham Market to Woodbridge, where it becomes navigable to small coasting

vessels and fishing-boats.   Here, too, it widens considerably
and becomes tidal.   Below Woodbridge it is joined by
two or three small tributaries, the chief of which is the
Fynn, which rises near Witnesham.   The Deben enters
the sea between Bawdsey and Felixstowe.

The River Orwell

Next we come to the Orwell, which above Ipswich
is known as the Gipping.   This is the most important
Suffolk river so far as navigation is concerned, for Ipswich
is a port with considerable maritime trade.   The Gipping
has its source near a village of that name, a few miles
north of Stowmarket, and before it reaches Ipswich it is

joined by several small tributaries. It is navigable to fairly large craft from Stowmarket downwards. The Orwell is really the estuary of the Gipping, and it is bordered by pleasant parks and woodlands and some picturesque villages. A service of steamers carries passengers on the Orwell between Ipswich and Harwich and Felixstowe.

The last of the eastward-flowing rivers is the Stour, which rises in the south-western part of the county and forms its southern boundary. The principal places on its Suffolk bank are Clare, Long Melford, and Sudbury, and it is navigable to barges as far up its course as the last-named town. Its principal Suffolk tributaries are the Glem, the Box and the Brett, and its waters, like those of the Orwell, enter the sea through Harwich Harbour.

As we have already learnt, the Little Ouse and the Lark, the rivers draining the north-western portion of Suffolk, are tributaries of the Ouse. The Little Ouse or Brandon River has its source at South Lopham, where the Waveney also arises, but while the latter flows in an easterly direction, the former flows towards the west. The only towns on its banks are Thetford (the greater part of which is in Norfolk) and Brandon. Below Brandon it enters the Fens and joins the Ouse. Its principal Suffolk tributary is a small stream which rises in the neighbourhood of Woolpit near Stowmarket, and is sometimes called the Upper Ouse.

The Lark, which has its source in some high ground near Stanningfield, is still a very small stream when it reaches Bury St Edmunds, where it is joined by the

Linnet.   It then flows to Icklingham, where it receives
the Kennet, and after forming the boundary of the
county for a few miles near Mildenhall it enters
Cambridgeshire.

Having learnt the names and courses of the principal
rivers of Suffolk, we will conclude this chapter with a
brief reference to two or three well-known lakes or
broads lying in the north-eastern portion of the county.
One of these, Oulton Broad, we have mentioned in
connection with the Waveney.   It provides some of the
best sailing that yachtsmen can enjoy in the Broads
district, and several regattas are held on it during the
Broadland season.

A few miles north-west of Oulton Broad is Fritton
Lake, a lovely wood-engirdled sheet of water of about
160 acres.   This is a private lake, but the public have
access to it during the summer for boating and fishing.
In the winter considerable numbers of wild-fowl are
caught on the lake by means of decoy-pipes.   The town
of Lowestoft is supplied with its water from Fritton Lake.

Quite close to the beach, between Kessingland and
Southwold, there are two small sheets of water known as
Covehithe and Easton Broads ; while adjoining the
Waveney, a few miles below Beccles, there is another
small broad, at Barnby.

Breydon Water, the estuary of the Norfolk river
Yare, forms the county boundary for a few miles.   It
is about 1000 acres in extent at high tide, but when the
tide is at ebb the water is confined to a narrow channel
bordered by wide mud-flats.

6. Geology and Soil.

By Geology we mean the study of the rocks, and we must at the outset explain that the term *rock* is used by the geologist without any reference to the hardness or compactness of the material to which the name is applied; thus he speaks of loose sand as a rock equally with a hard substance like granite.

Rocks are of two kinds, (1) those laid down mostly under water, (2) those due to the action of fire.

The first kind may be compared to sheets of paper one over the other. These sheets are called *beds*, and such beds are usually formed of sand (often containing pebbles), mud or clay, and limestone or mixtures of these materials. They are laid down as flat or nearly flat sheets, but may afterwards be tilted as the result of movement of the earth's crust, just as you may tilt sheets of paper, folding them into arches and troughs, by pressing them at either end. Again, we may find the tops of the folds so produced washed away as the result of the wearing action of rivers, glaciers and sea-waves upon them, as you might cut off the tops of the folds of the paper with a pair of shears. This has happened with the ancient beds forming parts of the earth's crust, and we therefore often find them tilted, with the upper parts removed.

The other kinds of rocks are known as igneous rocks, which have been melted under the action of heat and become solid on cooling. When in the molten state they have been poured out at the surface as the lava of volcanoes, or have been forced into other rocks and cooled in the cracks and other places of weakness. Much material is also thrown out of volcanoes as volcanic ash and dust, and is piled up on the sides of the volcano. Such ashy material may be arranged in beds, so that it partakes to some extent of the qualities of the two great rock groups.

The production of beds is of great importance to geologists, for by means of these beds we can classify the rocks according to age. If we take two sheets of paper, and lay one on the top of the other on a table, the upper one has been laid down after the other. Similarly with two beds, the upper is also the newer, and the newer will remain on the top after earth-movements, save in very exceptional cases which need not be regarded by us here, and for general purposes we may regard any bed or set of beds resting on any other in our own country as being the newer bed or set.

The movements which affect beds may occur at different times. One set of beds may be laid down flat, then thrown into folds by movement, the tops of the beds worn off, and another set of beds laid down upon the worn surface of the older beds, the edges of which will abut against the oldest of the new set of flatly deposited beds, which latter may in turn undergo disturbance and renewal of their upper portions.

Again, after the formation of the beds many changes may occur in them. They may become hardened, pebble-beds being changed into conglomerates, sands into sandstones, muds and clays into mudstones and shales, soft deposits of lime into limestone, and loose volcanic ashes into exceedingly hard rocks. They may also become cracked, and the cracks are often very regular, running in two directions at right angles one to the other. Such cracks are known as *joints*, and the joints are very important in affecting the physical geography of a district. Then, as the result of great pressure applied sideways, the rocks may be so changed that they can be split into thin slabs, which usually, though not necessarily, split along planes standing at high angles to the horizontal. Rocks affected in this way are known as *slates*.

If we could flatten out all the beds of England, and arrange them one over the other and bore a shaft through them, we should see them on the sides of the shaft, the newest appearing at the top and the oldest at the bottom, as shown in the figure. Such a shaft would have a depth of between 10,000 and 20,000 feet. The strata beds are divided into three great groups called Primary or Palaeozoic, Secondary or Mesozoic, and Tertiary or Cainozoic, and lowest of all are the oldest rocks of Britain, which form as it were the foundation stones on which the other rocks rest. These may be spoken of as the Pre-cambrian rocks. The three great groups are divided into minor divisions known as systems. The names of these systems are arranged in order in the figure with a very rough indication of their relative importance, though the divisions above the Eocene are made too thick, as otherwise they would hardly show in the figure. On the right hand side, the general characters of the rocks of each system are stated.

	Names of Systems		Characters of Rocks
TERTIARY	Recent & Pleistocene Pliocene		sands, superficial deposits
	Eocene		clays and sands chiefly
SECONDARY	Cretaceous		chalk at top sandstones, mud and clays below
	Jurassic		shales, sandstones and oolitic limestones
	Triassic		red sandstones and marls, gypsum and salt
PRIMARY	Permian		red sandstones & magnesian limestone
	Carboniferous		sandstones, shales and coals at top sandstones in middle limestone and shales below
	Devonian		red sandstones, shales, slates and limestones
	Silurian		sandstones and shales thin limestones
	Ordovician		shales, slates, sandstones and thin limestones
	Cambrian		slates and sandstones
	Pre-Cambrian		sandstones, slates and volcanic rocks

With these preliminary remarks we may now proceed to a brief account of the geology of the county.

The oldest geological deposit in Suffolk is the Chalk, which here forms part of a great band extending from Salisbury Plain to Norfolk. It consists of the Lower Chalk (which is without flints) and the Upper Chalk. The Lower Chalk can best be seen in the western part of the county, between Thetford and Newmarket, where it lies close to the surface. In the neighbourhood of Ipswich and Bury St Edmunds there are outcrops of the Upper Chalk, from which a variety of fossils, in the shape of sea urchins, shells, corals, and remains of fishes, have been obtained.

The Chalk, which is the only Secondary formation occurring in the county, represents a period when Eastern England and a considerable portion of Europe were covered by a wide sea, at the bottom of which the Chalk was deposited as a kind of ooze, chiefly made up of the shells of minute animals called foraminifera.

The oldest Tertiary deposits in Suffolk are the Thanet Beds, the best exposures of which are around Sudbury, in the Stour valley. They consist of clayey sand resting upon the Chalk and overlain by grey and pinkish sand.

Next in stratigraphical order we have the Reading Beds, of mottled clays and sands. These beds are worked for brick-making in several places in the southern part of the county.

Overlying the Reading Beds is the London Clay, which owes its name to the fact that it underlies the city of London. It can be seen in the cliffs at Bawdsey and in several places in that triangular tract of country lying between the estuaries of the Stour and the Orwell. At Woodbridge it rests upon a layer of white and yellow sand, from which many interesting fossils have been obtained, among them being portions of an extinct animal known as the Hyracotherium, a very early ancestor of the horse.

The Reading Beds and the London Clay represent a time when the climate of England was much warmer than it is to-day, and when palm trees grew and crocodiles and turtles lived in this part of the world.

Immediately overlying the London Clay is an interesting deposit called the Boxstone or Suffolk Bone Bed, in which the bones of various land mammals, including the mastodon, the rhinoceros, the tapir, and the hyaena have been found, as well as remains of sharks and whales. This deposit was formerly extensively worked for coprolites, which were used to improve the land for agricultural purposes. The "box-stones" are rounded masses of brownish sandstone that once formed part of a sandstone stratum which, in the days when there was a land connection between England and the Continent, extended across the area now covered by the North Sea.

The next deposits to claim our attention are the Crag beds. The lowest and oldest of these is the Coralline Crag, which is almost entirely composed of shells and shell fragments. The main mass of this crag occurs along the west side of the lower reaches of the Alde, between Aldeburgh and the Butley River, but there is also a smaller patch of it immediately north of the Alde. It probably represents a series of submarine banks which formerly fringed the western margin of the North Sea.

The Red Crag, of which the greater part of the cliffs at Felixstowe is formed, is another shelly deposit which has been divided by a well-known geologist into three

Section of Felixstowe Cliff showing the Red Crag

zones or periods, each possessing a distinct fauna. Two of these zones, the Butleyan and the Newbournian, are represented in this county, the third occurring in Essex. The Butleyan can best be seen at Butley and Bawdsey, and the Newbournian at Newbourn and Sutton. The

Red Crag, like the Coralline, came into existence as a submarine sand and shell bank.

The third Crag deposit occurring in Suffolk is the Norwich Crag, which is more recent than the Coralline and Red Crags, and which underlies the coast towns and villages between Leiston and Lowestoft.

Next we have the Chillesford Beds, consisting of thin layers of grey clay and white and brown sand. They can be seen in the cliffs between Kessingland and Lowestoft ; also at Southwold, Halesworth, and Beccles. They indicate that after the sea had retreated from the greater portion of Suffolk, the estuary of a great river, which probably flowed from the south-east, extended over part of East Suffolk.

Among the most interesting deposits of this county are the so-called Forest Bed series, which can be studied in the cliffs between Lowestoft and Kessingland when the waves have scoured away a good deal of the beach sand. These beds owe their name to their containing many trunks of trees, which were formerly supposed to mark the site of a buried forest; but it is now known that they were floated to the places where they are found by some great river, which left them stranded on its shores or mud-banks. Mr F. W. Harmer, our chief authority on East Anglian geology, believes this river to have been the Rhine; for at that time England was connected by land with the continent of Europe, and the Rhine probably flowed through a part of East Suffolk on its way to the sea. The bones of many huge and remarkable extinct animals have been discovered in the Forest Bed, among them being those of two kinds of elephant, the cave bear, and the sabre-toothed tiger. Three species of rhinoceros, the hippopotamus, and a hyaena also appear to have inhabited this part of England in the Forest Bed period.

In our study of the geological history of Suffolk, we now come to the Glacial period or Great Ice Age. During this period the northern and central parts of England must have looked very much as Greenland looks to-day. Intense cold prevailed, and large glaciers descended from the high ground and spread over the plains. One or more of these glaciers descended upon Suffolk, and when the ice melted it left behind it a great mass of clayey soil filled with fragments of various rocks over which the glaciers had passed. This clayey soil is called the Chalky Boulder Clay, because it contains a great deal of chalk. It extends over nearly the whole of Central Suffolk, and it can also be seen in cliff sections between Gorleston and Kessingland.

There are several different deposits in Suffolk belonging to the Glacial period, among them being the Contorted Drift, the Pebbly Series, and the Middle Glacial Sands. These can be studied in the cliffs between Gorleston and Kessingland, and geologists have concluded that they denote a long continuance of the Ice Age, but that between periods of intense cold there were milder intervals.

Overlying the Boulder Clay in East and South Suffolk and elsewhere, we have some beds of Plateau or Cannon-shot Gravel. These represent the time when the glacial ice was melting, causing floods which deposited the gravels.

The last geological deposits with which we have to deal are the Alluvial beds. The older of these are chiefly beds of gravel, deposited along the sides of valleys when the rivers were flowing at higher levels than they do to-day. They are of great interest to the antiquary because they contain flint implements made by man in prehistoric times. At a lower level we have the marsh and fen deposits, which, with the mud-flats of our estuaries and the sand-dunes along the coast, represent the latest-written chapter in the geological history of Suffolk.

A few words may be added about the soil of Suffolk. From what we have already learnt about the geology of the county, it is evident that the greater part of the surface soil consists of Chalky Boulder Clay, which forms the stiff heavy land on which most of the wheat is grown. Along the coast between Lowestoft and Kessingland this soil also occurs, but further south the coast lands are sandy or gravelly, and considerable tracts of it remain uncultivated. In the north-west a large extent of country has a surface soil of sand and gravel overlying the chalk. Here, too, there are large heaths and warrens, and very little corn is grown.

In the north-west there is also a small portion of the fen district, where the soil consists of clay, loam, and peat.

A good deal of marshy soil borders some of the rivers. As a rule this is not cultivated, for its lush grass provides excellent pasturage for cattle.

7. Natural History.

By studying the geology of Suffolk we have learnt
that at one time England formed a part of the continent
of Europe, with which it was connected by land extending
over a portion of the area now covered by the North Sea.
If it were possible to raise the whole of Great Britain and
Ireland, together with the bed of the sea and the coast of
Western Europe, about 600 feet, England and Europe
would be united again, while Ireland would be joined to
England, as it used to be.

When Great Britain and the Continent were joined
together, our country was inhabited by the same kinds of
animals as were to be found in Western Europe.   Many
of them are now extinct, including the sabre-toothed
tiger, the cave-bear, and others we have read about in the
geological chapter.   Most of the geologists attribute the
disappearance from England of these remarkable animals
to the severe cold and consequent sterility of the Glacial
period, but some think that they became extinct because
the whole of Great Britain, with the exception of the
peaks of the highest mountains, was for a time covered by
the sea.

Whatever may be the true explanation of their disappearance, there is no doubt that after the Glacial period
Great Britain was still united with the Continent, and
while this connection existed such animals as have inhabited our country since that time were able to reoccupy
it.   At the present time, however, Great Britain does not

possess so many species as France and Belgium, while Ireland has fewer species than England. The reason is that Great Britain became separated from the Continent before all the continental species had found their way here, while Ireland became separated from Britain before some of the British species, coming as they did from the south and east, had time to reach it. The same explanation accounts for Great Britain and Ireland having fewer wild plants than Western Europe.

The flora of Suffolk consists of a great number of species, and in consequence of the county having a variety of soils the wild flowers of some districts differ from those of others. Perhaps the most interesting district from the botanist's point of view is the north-west, where, on the heaths and warrens of Breckland, some sea-side plants are found, e.g. the dwarf-tufted centaury (*Erythræa littoralis*), the golden dock (*Rumex maritimus*), and the sand sedge (*Carex arenaria*). These are believed to be the survivors of a coast flora which grew there when an arm of the sea extended to the western border of Suffolk by way of the Wash and the Fens. Coast insects, too, are found there, e.g. *Agrotis valligera*, and the ringed plover (*Charadrius hiaticula*), which nowhere else nests away from the coast, always returns to Breckland in the nesting season.

As the greater part of the county has a clayey surface soil the wild plants suited to that kind of soil are the most abundant; but on the sandy lands bordering the east coast other plants are found, while in the western part we see the chalk-soil plants. Along the seashore the characteristic marine plants grow freely, among them being the very rare sea pea (*Lathyrus maritimus*) which is found near Aldeburgh. Salt-marsh flowers, sedges, and grasses occur in the marshy ground bordering the river estuaries, while higher up the rivers we get many beautiful marsh and riverside flowers.

Some idea of the floral richness of Suffolk may be gained when we learn that the county possesses 85 wild flowering plants which do not grow in more than 12 of the 112 counties and vice-counties into which England, Wales, and Scotland are divided by the botanists. Twenty-six orchids are included in the flora, some of them being very rare.

The list of the birds of Suffolk is a very interesting one, but unfortunately it includes several species which formerly nested in the county but, for one reason or another, no longer do so. Among these lost birds are the great bustard, the spoonbill, the avocet, the black-tailed godwit, the ruff and the kite.

The most interesting features of Suffolk bird-life may be briefly mentioned. In the neighbourhood of Aldeburgh there is a colony of common terns, these birds nesting in the sand and shingle of the seashore. Stone curlews come every summer to the heaths and warrens of the Breck district, where they are probably more numerous than anywhere else in Great Britain, and are hence generally called the Norfolk plover. Marsh birds are fairly numerous in the Suffolk portion of the Broads district, and wild-fowl come to the coast and estuaries in great numbers during sharp winter weather. At Fritton Lake near Lowestoft, and at Orwell Park near Ipswich, there are decoy-pipes for catching wild-fowl, many duck being captured in this way in some winters. These decoy-pipes are long tunnels of network covering dykes or channels connected with a lake or large pond. The wild-fowl are lured into them by means of decoy-ducks and dogs, and when they have entered a pipe the decoyman frightens them into a kind of hoop-net, from which they cannot escape.

Duck-Decoy at Orwell Park

Among the rare birds which have been observed in the county are Pallas's sand-grouse, the Greenland falcon, the Alpine swift, the cream-coloured courser, and the sooty tern.

Otters are not uncommon in the valleys of most of the Suffolk rivers, but the marten is extinct and also, in all probability, the polecat. The foxes, badgers, and dormice now inhabiting the county are said to have been recently imported, and not to be the indigenous races. Hares are plentiful in some districts, and rabbits so abound in Breckland that large tracts of warren called rabbit farms are hired solely that the rabbits may be killed for the market.

A rare species of mouse, known as the yellow-necked mouse (*Mus flavicollis*), has been caught in the parishes of Tostock, Whepstead, and Brettenham, and there are colonies of lesser or pygmy shrews (*Sorex pygmæus*) at Herringfleet and elsewhere.

Of the few Suffolk amphibians, the most interesting is the natterjack or running toad, a small toad which never hops.

8. The Coast. Landguard Point to Dunwich.

By carefully studying a map of Suffolk, we may soon become acquainted with the names and situation of its chief towns and villages; but to know the name of a place and where it is situated is not very interesting unless we have some idea of the kind of place it is and how it differs from other places. The coast-line of Suffolk, too, as we see it on a map, conveys no idea of its varied scenery; we cannot even tell whether the shore be flat or bordered by high cliffs. In this and the next chapter we will follow the coast-line of Suffolk from Landguard Point in the south to Gorleston in the north, noting as we go its most interesting features.

Landguard Point, which shelters the entrance to Harwich Harbour, is the most conspicuous promontory on the coast. Near the Point there is a lighthouse, and

Felixstowe Beach

close by stands Landguard Fort. There was a fort here as long ago as 1553, and in 1667 it was attacked by the Dutch, who succeeded in landing 3000 men, but they were repulsed and compelled to return to their ships. Early in the eighteenth century the fort was reconstructed, and in 1872 it was converted into a casemated battery for heavy guns.

A two-mile walk over common-land brings us into Felixstowe, one of the most popular watering-places on the coast. Here the low flat land of the Point gives way to cliffs and cliff gardens, the older part of the town being irregularly built along the sea front. The town takes its name from Felix of Burgundy, who introduced Christianity into East Anglia, and who is believed to have had a "stowe" or dwelling here.

The cliffs here consist chiefly of London Clay at the base, overlain by the Red Crag, which is largely made up of shells. This Crag caps the cliffs the greater part of the way to Bawdsey Haven, the mouth of the river Deben; but as the river is approached the cliffs gradually become lower, and disappear before we reach the Felixstowe golf-links.

There is a ferry across the Deben at Bawdsey, landing us in a sparsely populated parish about five miles in length. It extends along the coast as far as Hollesley, where we come to the entrance to Orford Haven. Here, in order to keep to the coast-line, we must cross the river Alde or Ore, which for several miles flows parallel with the coast, separated from the sea by a narrow strip of sand and shingle. The beach here is lonely and desolate, in consequence of which it is frequented by many sea and shore birds, including hundreds of terns in the summer.

We find the coast inhabited again as we approach Orford, a decayed little town that was formerly important enough to return two members to Parliament. A strongly fortified castle, of which the massive keep remains, was built here about 1165.

Another lonely tract of sand and shingle stretches from Orford to Slaughden, a hamlet forming part of the town of Aldeburgh. Here the river Alde lies within about a hundred yards of the sea, and it was on Slaughden

The Beach at Slaughden
(Showing cottages partly destroyed by the sea)

Quay, bordering the river, that George Crabbe, the poet, who was born at Aldeburgh, worked as a quay-labourer when he was a young man. Just before we enter Slaughden, we see one of the little forts called Martello towers, which were erected early in the nineteenth

century when a French invasion of our coast was
expected.

Slaughden is a little sea-wasted hamlet, where most of
the few pebble-built cottages have been destroyed by the
waves; but Aldeburgh, although it has been encroached
upon by the sea, is a rising watering-place, built partly on
low ground bordering the beach and partly on rising ground
a short distance inland. This town, too, was formerly a
Parliamentary borough returning two members. In its
immediate neighbourhood the country consists largely of
heathland and salt marsh. The characteristic features of
its scenery have been graphically and accurately described
by Crabbe in his poems.

Northward of Aldeburgh a road known as the Crag
path runs beside the beach to Thorpe, a small hamlet in
which is Thorpe Ness. Beyond the Ness is another
sparsely populated stretch of low-lying coast. The land
rises, however, towards Dunwich, which is approached
across a tract of furzy heathland, bordered by sandy cliffs.

Dunwich is in some respects the most interesting place
on the Suffolk coast. Now only a small village, it was
once the largest town in the county, possessing several
monastic houses, hospitals, and churches. From an old
MS. preserved in the British Museum, it appears that it
was strongly fortified; for we are told that in the reign of
Henry II, when Robert Earl of Leicester tried to take the
town, "the strength thereof it was terror and fear unto
him to behold it, and so retired both he and his people."

The sea destroyed ancient Dunwich, and the remains
of most of its churches and monasteries are strewn about

the ocean bed. In the fourteenth century, its historian records, 400 houses were destroyed, and by the middle of the sixteenth century four of its churches were washed away. Early in the eighteenth century another church went "down cliff," and at the present time the ruins of All Saints Church are standing on the verge of the cliff (*v.* Fig. on p. 45). This ruined church, with the ivy-clad fragments of a Franciscan friary and some remains of a Norman chapel, are all that is left of a town which from 1296 until 1832 returned two members to Parliament and which was a corporate town until 1886.

9. The Coast. Dunwich to Gorleston.

Northward of Dunwich there is a wider beach of sand and shingle on which the sea poppy and the sea spurge grow; and this beach extends to the borders of the picturesque coast village of Walberswick, a favourite haunt of artists during the summer and autumn months.

This village borders the river Blyth, which can be crossed here by means of a ferry. On the north side of the river is Southwold, one of the most popular of Suffolk watering-places. It stands on fairly high ground, with a breezy common on the south and some very picturesque scenery in its immediate neighbourhood.

It was in Sole Bay, off Southwold, that a great naval battle was fought in 1672 between the allied fleets of England and France under the Duke of York and Admiral D'Estrees and the Dutch fleet under De Ruyter. The allied fleet numbered 101 ships, while De Ruyter commanded 91 men-of-war, 54 fire-ships and 23 tenders. In this sea-fight the French fleet took only a small part. The English lost six ships, including the *Royal James*, the flag-ship of the Earl of Sandwich; while the Dutch lost three ships and their Vice-Admiral van Ghent.

Southwold new Harbour and Quay

Some old-fashioned cannon standing on what is called the Gun Hill, at Southwold, are a reminder of a famous land battle. They were taken at Culloden by the Duke of Cumberland, who gave them to the town.

Adjoining Southwold is the parish of Easton, which could once boast of having in Easton Ness the most easterly land in England, but the Ness has been long ago washed away by the sea. A small sheet of water called Easton Broad lies close to the beach, and a short distance further northward there is another small broad, known as Covehithe Broad.

An interesting geological feature of the coast at Covehithe is a submerged peat bed, representing a tract of marshy land which formerly bordered this part of the coast. When the tide is low, portions of this peat bed are to be seen on the edge of the beach, frequently with the stumps of small trees and bushes still rooted in them. After stormy weather raft-like masses of peat are occasionally broken off the mass at the bottom of the sea and strewn along the shore.

The church at Covehithe is, perhaps, the most picturesque on the Suffolk coast. It consists of the ruins of a large and fine old church, within the nave of which a much smaller and more modern church has been built.

A fine wide expanse of beach lies between Covehithe and Kessingland, bordered on its landward side by undulating sand-dunes. A small river, draining a tract of marshland, seems to have entered the sea here at one time; but now the drainage of the marshes is effected by means of dykes and a sluice.

Kessingland is a growing village having its share of the holiday-making visitors to the coast. It is the point where a line of cliffs begins, which extends as far as Lowestoft. These cliffs, which in places rise to a height of about 70 feet, consist chiefly of glacial sand, upon which rests a considerable mass of chalky boulder clay, while beneath the sand are the Chillesford and Forest Bed deposits described in our geological chapter. One or two clefts or "gaps" in the cliffs were formerly much larger than they are to-day, and they are said to have been resorted to by smugglers who landed cargoes on this part of the coast.

Lowestoft, the largest town in the county with the exception of Ipswich, the county town, includes the parish of Kirkley. It is one of the principal watering-places of England and also one of the chief fishing ports. Its important fisheries will be described in a later chapter. The older portion of the town is built on the high ground of a cliff from which the sea long ago receded, at the base of which is the fishermen's quarter, bordering the beach. The principal hotels and boarding houses for visitors are on the south side of the harbour, where there are two piers and a fine esplanade; but northward of the town, where the old line of cliffs extends towards Corton, many houses have been built and more are likely to be erected. A lighthouse called the Low Light stands on the Ness, the most easterly point of England, and another light-house, called the High Light, is built on the cliff, near the site of an old beacon.

From the Low lighthouse a pleasant expanse of grass-grown sandy land called the Denes extends northwards to Corton, bordered on its landward side by cliff slopes which in late summer are purple with heather. Corton is a fishing village, popular with summer visitors. Beyond it there are sandy cliffs again, extending in an unbroken line to Gorleston.

This last-named town is a rising watering-place situated

on the south side of Yarmouth Harbour, through which the waters of the Yare and of the other rivers of the Broads district make their way to the sea. It is the northernmost parish on the Suffolk coast, so having reached it our coast journey comes to an end.

10. The Coast. Lost Land.

In several places along the coasts of England large tracts of land have been won from the sea within historical times. For instance in Kent the land known as Romney Marsh was formerly almost entirely covered by the sea, while north and south of King's Lynn, in Norfolk, thousands of acres of salt marsh have been embanked and reclaimed during the last hundred years. Both in Kent and Norfolk there are towns which were once coast towns, but which are now situated some miles inland.

Along the Suffolk coast there is hardly a spot where any such gain of land has been effected, either by natural or artificial means. Indeed, it may be said that the story of the coast is almost entirely one of destruction and sea encroachment. Precisely how great the loss of land has been during historical times, it is impossible to ascertain; but we know that for many centuries the sea has been gaining on the land and that in several places it is still gaining. From Felixstowe in the south to Lowestoft in the north there is not a coast town which has not suffered seriously in consequence of the inroads made upon it by the sea.

Some coasts, protected by bold and hard rocks, are affected very slightly by the beating and scouring of the waves, but the Suffolk coast can offer little resistance when high tides occur in stormy weather. Its cliffs, as we have already learnt, consist for the most part of sand, and between the lines of cliffs the shore lies low and exposed to the full force of the breaking waves.

We will try to get some idea of the changes that have taken place along the coast.

Returning to Felixstowe, we find that here the greater part of a parish called Walton has been washed away since the time of the Romans, whose strong fortress has quite disappeared. A small harbour called Wadgate is recorded to have existed at Felixstowe in the reign of Edward III; but of this, too, there is now no trace.

In the neighbourhood of Orford the coast has undergone considerable alteration. At one time the river Alde must have entered the sea some miles northward of its present mouth, but the gradual extension southward of the narrow strip of land lying between the river and the sea has forced the river to flow parallel with the coast for several miles before it enters Hollesley Bay.

The little town of Aldeburgh now consists chiefly of two streets running nearly parallel with the beach. Formerly there were three such streets; but one has been entirely destroyed with the exception of an ancient Moot Hall, which now stands a few yards from the sea. The adjoining hamlet of Slaughden, too, has suffered in a similar way, and the broken walls of ruined houses can be seen half embedded in the shingle of the beach.

In our last chapter we referred to the destruction of the once important town of Dunwich. There is a tradition that a forest called Eastwood formerly lay between the town and the sea, but this tradition may have arisen in consequence of the existence of traces of a submerged peat bed, like that to be seen at Covehithe. An old

Ruins of All Saints Church, Dunwich, at edge of Cliff

record describes the town as having been " of great force, and strong enough to keep out a great number of people," and there is no doubt that it had several monastic houses and at least a dozen churches. Of its harbour there is now no trace, and as the sea still continues to undermine its cliffs it will not be long before the last of its old churches is totally destroyed.

At Southwold the sea has encroached considerably in recent years, notwithstanding that groynes and break-waters have been constructed to protect the cliffs. This recent damage seems to be a repetition of early history, for in 1619 an appeal was made to the country on behalf of the Southwold fishermen, whose harbour had suffered through "the violence of the water."

A little way north of Southwold, a very small tract of land is all that remains of the parish of Easton, which in the reign of Edward III was a place of some importance, possessing a market. Its church of St Nicholas is believed to have been destroyed by the sea.

Covehithe, too, the next parish, has had its area considerably reduced by the wasting of its cliffs. At times, when portions of the cliffs have fallen, relics of the Roman period have been discovered, indicating that there was a settlement of the Romans here. In an earlier period a tract of fen or marshland extended some distance seaward of the present coast-line.

Between Covehithe and Kessingland the beach is bordered by sandhills, and the sea does not appear to have gained upon the land to any considerable extent in recent years; but northward of Kessingland entire fields have disappeared within the memory of men still living. Almost every year large masses of cliff fall on to the beach, and because of this a newly erected Naval Reserve Battery at Pakefield had to be pulled down a few years ago. Pakefield, in spite of attempts at protection, has suffered severely during the last fifteen years. Several houses have "gone down-cliff" and more are likely to follow them.

At Lowestoft a hard fight against the sea has been kept up for several years, and during the last two or three years the sea has been kept from gaining further upon the

Coast Erosion at Pakefield

land. Northward of the harbour, where there is a tract of level ground from which the sea receded many centuries ago, the Low Lighthouse has twice been removed on account of its being threatened by the sea.

Between Lowestoft and Gorleston there has been sea encroachment at Corton, and a hamlet called Newton has entirely disappeared. Great changes must have taken place along this part of the coast; for until the early part of the fourteenth century the entrance to Yarmouth Harbour was opposite the parish of Corton.

Where the coast is protected, it is chiefly by means of sea-walls and groynes. At Lowestoft large sums of money have been spent in providing such protection. Northward of the town a long sea-wall safeguards a pleasant expanse of open denes, while the cliffs at Kirkley, a parish within the bounds of the borough, are also faced by a concrete wall. At Corton a massive sea-wall has been built in front of a large private estate; while at Lowestoft, Southwold, Corton and elsewhere wooden groynes have been constructed to cause an accumulation of sand and shingle along the beach.

11. The Coast. Sand=banks and Light-houses.

The Suffolk coast is rather dangerous to shipping, and a winter never passes when the services of the lifeboats are not required at several stations between the mouths of the Orwell and the Yare. Projecting far into the North Sea, this easternmost coast of England is skirted by many thousands of coasting ships every year, in the course of their voyaging between the northern and southern ports; while the fishing vessels belonging to the port of Lowestoft are continually passing in and out of that harbour. Consequently, in stormy or foggy weather it frequently occurs that several vessels run ashore or upon some of the sand-banks lying off the coast.

But the loss of ships and lives along this coast has not been nearly so great of late years as it was when the coasting trade was entirely carried on by sailing ships. On December 18, 1770, a storm occurred early in the morning, and at daybreak 18 ships were seen on a sand-bank off Lowestoft, half of which number went to pieces before nine o'clock. Many other ships were sunk close by, and it was estimated that about 200 sailors were drowned.

Off the southernmost part of the coast there are several small sand-banks, among them being the Cork Ledge and the Cork Knoll. Opposite Bawdsey is the Cutter Sand, seaward of which lie the Kettle Bottom, Bawdsey Bank, and the Shipwash. Between Orford Ness and Thorpe Ness are the Ridge and the Gabbard; while a little way north of Thorpe Ness is the Sizewell Bank. Further northward, too, there are some large sand-banks, notably the Barnard, off Kessingland, and the Newcome and the Holme, off Lowestoft; while between Corton and Gorleston lies the Corton Sand.

The presence of so many sand-banks makes it necessary that the buoyage system should be very complete. Sixty-two buoys are placed between Harwich Harbour and the mouth of the Yare, some of them being lighted with gas. Buoys frequently have to be moved in consequence of the shifting of the sands.

In addition to the buoys, there are four lightships stationed between Harwich and Gorleston. They are named the Cork, the Shipwash, the Outer Gabbard, and the Corton lightships. Each ship has a red hull with the name painted on both sides; it also bears a distinguishing mark in the shape of a ball, a cone, or the upper half of a ball at the mast-head; and the light is either white or red. These lights either revolve or flash at intervals, and they are visible 10 or 11 miles. In foggy weather fog-horns or sirens are sounded to warn ships of danger.

Orford Lighthouse

England has taken the lead in seafaring ever since the reign of Elizabeth, but until about a hundred years ago little was done in the direction of lighting the coast for the guidance of mariners. Up to that time the method of coast-lighting was the beacon fire, which was kept burning on the top of a stone or brick tower, on a high building, or on a hill near the sea. At Orford a beacon fire was kept burning on the top of the castle keep, and at Lowestoft there was a beacon tower, the foundations of which can still be seen, near the present High Lighthouse.

The last of the English beacons was not extinguished until 1822. Since the beginning of the nineteenth century nearly 900 lighthouses have been erected round the British coasts.

The principal Suffolk lighthouses are at Orford, Southwold, and Lowestoft. The Orford light on the Ness is an occulting light of white, red, and green, visible 15 miles. Once every 40 seconds it disappears for 3 seconds. A lower fixed white light on the same tower is visible 6 miles. At Southwold there is what is called a group occulting light of white and red sectors, visible 17 miles.

At Lowestoft there are high and low lighthouses. The former, standing on high ground at the north end of the town, is a white light revolving every half minute and visible 17 miles. In the same tower, below the main light, there is a fixed red sector. The low lighthouse on the Ness shows an occulting light of red and white sectors visible 14 miles. This light disappears for 3 seconds every half minute.

In addition to the lighthouses, there are several smaller lights marking harbour mouths, pier-heads, and channels between sand-banks.

12. Climate and Rainfall.

The climate or average weather of a country has a marked influence upon life in that country, both animal and vegetable.   England, as we know, has a temperate climate, but even in England the climate of one county often differs considerably from that of another.   Situation and height above sea level are chiefly responsible for this difference.   In the case of Suffolk, the climate is influenced by the nearness of the sea and the character of the surface soil, but other factors come into play.   Towns and villages occupying sheltered positions are not so exposed to cold winds as are those with a northerly or easterly aspect; while places situated on a southerly slope of the land are naturally more sunny than those facing the north.   The amount of rainfall, too, is influenced by the configuration of the land and the character of the vegetation.   Rain-clouds passing over a flat country will often separate when they come to a tract of higher ground, so as to leave it dry; while it is well known that woodland areas usually have a heavier rainfall than treeless districts.

When we compare the temperature of Suffolk with that of other parts of England we get some interesting and instructive results.   For the purposes of this com-parison, we will take the year 1907.   In that year the mean temperature of Lowestoft was 48·2° Fahr., of Felix-stowe 48·9°, and of Geldeston, near Beccles, 48·4°.   These figures compare very favourably with those of most places in the Midland counties; but they are rather lower than

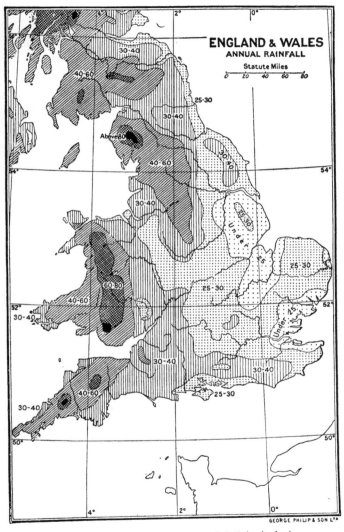

ENGLAND & WALES
ANNUAL RAINFALL
Statute Miles
0 20 40 60 80

30-40
40-60
25-30
30-40
Above 80
40-60
30-40
30-40
Under 25
25-30
40-60
60-80
40-60
30-40
25-30
Under 25
30-40
30-40
30-40
40-60
30-40
25-30

GEORGE PHILIP & SON LTD

(The figures give the annual rainfall in inches.)

those of places like Falmouth and Plymouth, in the south-west of England, where the mean temperature was 50·6° and 50·3° respectively. When, however, we take the records for three years (1905–1907), we find that Felixstowe, owing to its sheltered position, has an average mean temperature of 49·4°, which is only about one degree lower than those of Brighton and Bournemouth.

We may now turn to the rainfall. The map over-leaf shows very clearly that, speaking generally, the rainfall of England decreases steadily as we pass from West to East. The moisture-laden clouds coming with the pre-valent winds across the Atlantic condense on the higher lands in the West, and in the country beyond them there is consequently less rain. Some parts of Suffolk fall within the driest area of England, where the rainfall is under 25 inches. In 1907 the total rainfall at Lowestoft was 21·4 inches, at Geldeston 22·5 inches, and at Bury St Edmunds, in the western part of the county, 25·5 inches. At Felixstowe it averages about 20·8 inches.

At Laudale in the west of Scotland, on the other hand, the rainfall in 1907 was 76·8 inches, while at Glencarron in the north of Scotland it was 86·6 inches. In some places, owing to exceptional conditions, there is an even heavier rainfall, that of the neighbourhood of Snowdon being every year the heaviest in England. The average rainfall of England as a whole is as nearly as possible 32 inches. At Lowestoft, in 1907, rain fell on 172 days, and at Geldeston and Bury St Edmunds on 211 days.

In the matter of bright sunshine, too, Suffolk is eminently favoured. At Lowestoft, in 1907, there were

1719 hours of bright sunshine, and at Felixstowe 1726 hours. At Manchester, in the same year, only 894 hours of bright sunshine were recorded, while at Westminster there were 1234 hours.

Land fogs are rare in most parts of Suffolk; but in the autumn they are of frequent occurrence in the marshy districts, where on summer nights a good deal of mist rises from the rivers and dykes. Sea fogs of considerable density occur at times along the coast.

The prevailing winds are westerly, north-westerly, and south-westerly. At Lowestoft, in 1907, the wind blew from a westerly direction on 168 days. The east wind is usually experienced during February and March, when it often blows with considerable strength and keenness.

13. People—Race, Dialect, Settlements, Population.

When the Romans conquered Britain, Suffolk was inhabited by a people whom Julius Caesar calls the Cenimagni, but who were known a century later as the Iceni. Probably the majority of them were Brythons; but among them there may have been earlier settlers, called Goidels or Gaels, and there seems to be some evidence that representatives of an even earlier, and quite distinct, people, probably of Iberian stock, were still to be found in some parts of the county.

In north-west Suffolk, at the present time, many of

the inhabitants contrast strikingly with the rest of the people of the county, having very dark hair and dark eyes. At Brandon, where the ancient industry of flint-knapping seems to have been carried on ever since pre-historic times, several of the knappers represent this distinct racial type, and photographs of them have been mistaken for those of Welsh people. The same type occurs in some parts of the Fen district. These dark-eyed, dark-haired people are probably descendants of a race that inhabited Suffolk in the Stone Age, a prehistoric period when the inhabitants of our land had no know-ledge of the use of metals and made their tools and weapons of stone.

After the Romans had withdrawn from Britain, the Saxons and Angles came and settled in Suffolk, and it is upon the language spoken by them that our English tongue is based. The inhabitants of Suffolk to-day are chiefly descendants of the Anglo-Saxons. In this county, however, a good many words derived from the Dutch, and a few derived from the Norse, are in use. The Norse words probably came to us with the Northmen or Danes, while the Dutch words seem to have been intro-duced by later settlers, of whom something will be said presently.

It is not quite certain when the Northmen first estab-lished themselves in Suffolk. Many of them arrived during the Saxon period, but some antiquaries believe that there was a settlement of Northmen in the county before the Romans came. Professor Windle writes: "Perhaps the most important Danish contribution to place names is the suffix *by*. *By* or *byr* originally denoted a single dwelling, or a single farm....By degrees, like the suffixes *ton* and *ham*, it came to have a larger meaning and denoted a village."

Just beyond the border of Suffolk, in the county of Norfolk, there is a cluster of parishes each having a name ending with *by*, and these parishes were undoubtedly Danish settlements. Ashby and Barnby in the north-eastern corner of Suffolk are also, in all probability, parishes taking their names from settlements of the sea-rovers, the presence of whom is suggested by many place names on and near the coast. Among the Suffolk fishermen of the present day tall fair-haired blue-eyed men closely resembling the Norwegians are frequently met with.

Many of the Dutch words in use in Suffolk were probably brought to us by Dutch weavers, large numbers of whom settled in the county in the fourteenth and sixteenth centuries. Towards the end of the seventeenth century a considerable number of French Huguenots, driven from France by the revocation of the Edict of Nantes, made their homes in this county, where some of their descendants still reside.

Having learnt something about the origin of the greater part of the inhabitants of Suffolk, we will now come down to the present time and ascertain how many people inhabit the county, how many houses they live in, and a few other facts about them, which, with many more, are given in the published returns of the last census, taken in 1901.

In 1901 the county had a population of 384,293 persons. In 1801 the population was only 214,404, so it will be seen that a large increase has taken place during a hundred years. At the time of the last census considerably more than one half of the people were living in rural districts; but during the previous ten years the rural population had decreased about 10,000, while the population of the urban districts (including Ipswich, Lowestoft, and other towns) had increased by about 21,000. The decrease in the rural districts was mainly due to many of the sons and daughters of agricultural labourers having gone to work in towns and cities.

The census also shows that in Suffolk the females exceeded the males by 9,661. This means that there were 1053 females to every 1000 males. Only 498 people of foreign origin were living in the county, and more than half of these were German, French, or Italian.

From the census returns we also learn that the number of inhabited houses in the county was 83,788. There were 1043 persons in military or naval barracks or on board His Majesty's ships in the harbours; and 3,747 persons, excluding officials, occupying workhouses, hospitals, asylums, and reformatories. On board ships in the harbours and barges and wherries on the rivers there were 788 persons.

As Suffolk is chiefly an agricultural county it is interesting to know that in 1901 there were 3,747 farmers and graziers in the county, also 299 women engaged in managing farms. These found employment for 27,319 agricultural labourers, horsemen, and cattle-tenders.

The number of people to the square mile in Suffolk is 258. The average per square mile for the whole of

England and Wales is 558, or more than twice as great. Suffolk may thus be considered a sparsely inhabited county. In 1801 the population was 144 to the square mile, and in 1851 it was 226.

14. Agriculture—Main Cultivations, Woodlands, Stock.

We have already seen that Suffolk is an agricultural county. In view of the fact that the welfare of its inhabitants is so largely dependent upon agriculture, it will be interesting to learn what are its chief agricultural products, and how much land is devoted to their cultivation. From a report which is issued every year by the Board of Agriculture, we learn that of the 56,200,000 acres of land-area in Great Britain, 32,243,447 acres were under cultivation in 1907. Suffolk has a land-area of 945,411 acres, of which 756,017 acres, or about a third of the whole, consisted of arable land and permanent pasture.

The corn crops of Suffolk in 1907 consisted of wheat, barley, oats, rye, beans, and peas, which were grown on 341,582 acres. This means that nearly a third of the county was given up to the cultivation of corn crops. Barley was the largest crop, covering 127,020 acres, wheat being next with 96,375 acres. Oats, beans, peas, and rye followed in the order in which they are named.

The principal root and green crops are turnips and swedes, mangold, clover, sainfoin, vetches, and potatoes.

These covered 191,297 acres, or a little more than a fifth part of the county. Turnips and swedes constituted the chief root crop, being grown on 42,860 acres. Clover, sainfoin, lucerne, vetches, and grasses under rotation covered 108,822 acres, or about a ninth part of the land area.

Staverton Park

Formerly hops were grown in some parts of central Suffolk, but during the last few years the cultivation of this crop has been abandoned. In 1906 there were only two acres of hops in the county, and in 1907 none at all were grown.

Suffolk can hardly be ranked among the fruit-growing counties, for in 1907 only 835 acres were devoted to the growth of small fruits, such as strawberries, raspberries, currants, and gooseberries. Orchards of apple, pear, plum, and cherry trees covered 2,196 acres, but of this land nearly 300 acres were also used for the cultivation of small fruits.

There are no large woods in Suffolk to-day, although many years ago a considerable portion of Mid Suffolk was covered by ancient woodlands. The extent of land now covered by coppice, plantations, and other woods is about 38,000 acres, an increase of about 3000 acres since 1895. About 24,000 acres consist of permanent woodland, the remainder being coppice—or woods which are cut over periodically and which reproduce themselves by shoots from the base—and plantations made within the last ten years. The amount of timber grown cannot be accurately estimated from the above figures, for in almost every part of the county oak, ash, black poplar, elm, and other trees grow freely by the roadsides and along the field borders. Some of the oaks growing on the clay lands are especially fine old trees. The Great Oak at Cretingham is so large that its hollow trunk has been used as a shelter for stock and as a cart-shed. Queen Elizabeth's Oak, in Huntingfield Park, measured nearly 33 feet in girth at the height of 7 feet from the ground, but it is now in an advanced state of decay. In Henham Park is another venerable oak in which a Royalist member of the Rous family is said to have been concealed in the time of the Commonwealth.

The rearing of horses, cattle, sheep, and pigs is so much a part of the farmer's occupation that some reference to the live stock of Suffolk may be included in this chapter. Naturally, most of the horses are used in agricultural operations; consequently we find that out of 41,221 horses in the county in 1907 no fewer than 32,186 were

Prize Suffolk Ewe Lambs

(Bred by and the property of Mr H. E. Smith, Walton Grange, Suffolk)

employed for such purposes. The cattle in the same year numbered 77,013, the sheep 339,389, and the pigs 158,101.

Most parts of Great Britain have produced their own peculiar races of domesticated animals, and in this respect Suffolk is no exception. The Suffolk horse is one of the most famous breeds in England. It was formerly called the "Suffolk Punch" and is still known by this name abroad. A special society exists to maintain the purity of the breed. In 1618 a Suffolk writer described the Suffolk horse as being "among the many ornaments of this shire," and since then it has been greatly improved.

The Red Polled cattle of Suffolk, too, are widely famous. This fine breed was established about 1782. The sheep especially associated with this county is the black-faced sheep, the ewes of which have been in great demand for the production of the celebrated cross-bred Suffolk and Down. Suffolk is also noted for its large black pigs.

15. Industries and Manufactures.

Suffolk, although mainly an agricultural county, has several important industries and manufactures. Some of these are directly or indirectly connected with agriculture; others are largely dependent on the great herring and trawl fisheries; while yet others exist for which Suffolk was early famous.

One of the principal industries connected with agriculture is the making of farming implements, which is carried on largely at Ipswich, Bury St Edmunds, and Leiston. Ploughs, drills, harrows, threshing machines, and all kinds of steam and electrical engines are made at the large Suffolk iron-works and foundries, and they are exported to all parts of the world. Bury St Edmunds is

Boat-building yard at Lowestoft

noted for its manufacture of malting and brewing machines; while at Ipswich the making of large cranes and sluices is a speciality. At Bramford near Ipswich, at Stowmarket, and elsewhere there are extensive artificial manure works.

Many years ago Suffolk was noted for its ship-building. Beside the tidal waters of the Orwell and the Deben there were yards from which many ships were launched, among them being frigates carrying 40 guns. Ship-building in Suffolk came to an end when steamships were invented; but the loss thus sustained by the county has been more than made up to it by the great increase in the number of vessels engaged in the trawling and herring fisheries. These vessels, many of which are steamboats, are nearly all built at Lowestoft, the principal fishing port of the county.

Suffolk once had a great reputation for the manufacture of woollen goods. Cloth-weaving was introduced into the county by Dutch and Flemish weavers, the earliest of these settlers arriving in the fourteenth century. Hadleigh, Sudbury, and Lavenham were important centres of this manufacture, and in those towns some of the old houses that were occupied by the foreign weavers are still standing. Kersey and Lindsey were also among the weaving centres, and gave their names to kerseymere and linsey-woolsey, two fabrics which in all probability were first made there. At the present time the woollen industry is greatly decayed; but in some towns its place has been taken by the manufacture of other woven goods, including cotton, silk, linen, horsehair-seating, and coco-nut-matting. At Lavenham horsehair-weaving is carried on by a considerable number of women in their own homes. Sudbury and Glemsford are centres of mat-making and silk-weaving, and mat-making is also carried on at Hadleigh. In Haverhill the principal industry is

A Brandon Flint-Knapper

the manufacture of drabbet, huckaback, and other textile fabrics.

Brick-making is an industry established in many places. Woolpit, near Stowmarket, is famous for its white bricks, reputed to be the best in the world. At Elmswell the making of cricket bats is probably unique as a village industry.

At Stowmarket there are large works for the manufacture of guncotton, cordite, and other explosives.

At Bungay and Beccles many hands are employed in printing, the printing works of these towns being old-established and largely concerned in the printing of books for London publishers.

The most ancient Suffolk industry is flint-knapping, which may have been carried on in the north-western part of the county ever since prehistoric times. Brandon is the home of the Suffolk flint-knappers, whose work chiefly consists in making gun-flints for export to Africa and elsewhere, but they also undertake the shaping of flints for the flint-panelling of church walls. In the days when fire was obtained by the use of a flint and steel and tinder-box, Brandon men made most of the flint "strike-a-lights," and in recent times thousands of flints and steels were made for the use of our soldiers engaged in the war in South Africa.

16. Minerals.

Suffolk can put forward no claim to be considered a mining county, for it has no mines. In prehistoric times its inhabitants may have excavated flint from the chalk, as they did in the adjoining county of Norfolk, and as the Brandon flint-workers do to-day. On Lingheath, near Brandon, there are hundreds of pits which have been dug by the Brandon men, and it is interesting to know that these men use a single-pronged pick which in form closely resembles the picks made of deer's antlers used by the Stone Age men. The flint obtained from the chalk at Brandon is said to be the best in England.

In some parts of the county chalk is quarried for the making of lime and whitening, while hard chalk is ex-

Lingheath Flint Pits, Brandon

cavated for building purposes. The "clunch" extensively used in local church-building is a hard kind of chalk.

Several geological deposits in the county have been worked in order to obtain material suitable for improving the surface soil for agriculture. From the tract of land between the rivers Alde and Orwell phosphatic nodules known as coprolites, found at the base of the Red Crag, have been excavated; these nodules were crushed and ground to a powder used in the preparation of artificial manure for turnips and other root crops. For nearly two hundred years the Suffolk Crag itself has been used for improving the quality of the soil, while dressings of London Clay and Chalky Boulder Clay have been utilised for the same purpose.

In many parts of the county old "marl pits" are to be seen, often in the middle of fields. This marl consists chiefly of the chalk which is mixed up with the Chalky Boulder Clay. Large quantities of it have been dug up and spread over the light lands, which have been greatly improved by it. Occasionally it is sufficiently pure to be burned for lime.

The Reading Beds, described in our geological chapter, consist of clay and sand which is used for brick-making in the southern part of the county. London Clay and Boulder Clay are also employed for this purpose. Made into blocks and mixed with straw, the Boulder Clay is used in building clay-lump cottages.

In the London Clay there are layers of nodules, called "cement stones" in consequence of their having been much sought after by makers of cement. Off the Felixstowe coast these cement stones form what is called the West Rocks. From these rocks the stones have been obtained by dredging, with the result that a useful natural breakwater has been almost entirely removed.

Glacial and river gravels are extensively worked for road-making and road-mending, while in the neighbourhood of Ipswich the "box-stones" of the Suffolk Bone Bed have been used for mending the roads.

17. The Trawl and Herring Fisheries. Minor Fisheries.

It is natural that Great Britain, an island country, should have extensive sea fisheries, and that they should be mainly carried on by the inhabitants of towns and villages along the coasts. Sea fish forms a considerable portion of the food consumed by the people of this country, and the extent to which sea fishes are caught will be understood when it is stated that in the year 1905 no less than 11,365,000 cwts. of fish (not including shell-fish) were landed by our fishing vessels, their value being £6,490,000.

The British sea fishes which have a marketable value number about fifty, but only about thirty of these are caught in considerable quantities. They are divided into two classes. Some are called *demersal*, because they live and feed near the bottom of the sea; others are described as *pelagic*, because they swim about in shoals at or near the surface of the sea. As examples of the former class we will take the sole and the plaice, while the latter class shall be represented by the herring and the mackerel.

Now it is evident that, as the habits of the sole and the plaice differ so greatly from those of the herring and the mackerel, different methods must be used in capturing them. Consequently we find that the former are chiefly taken by trawling vessels, which drag their nets along the bottom of the sea, while the latter are taken by "drifters," which spread their nets just beneath the surface of the sea. In addition to this, a comparatively small quantity of demersal fish is taken by liners, i.e. vessels fishing by means of lines of baited hooks.

It may here be mentioned that the principal British fishing grounds are in the North Sea. Generally about six times as many demersal fish and about fifteen times as many pelagic fish are landed every year at the east coast ports as at ports on the west coast.

Lowestoft Trawlers leaving port

Lowestoft is the principal fishing port on the Suffolk coast. It carries on an important trawl fishery, its fleet of steam and sailing trawlers being one of the finest in the world.

From the Report on the British Sea Fisheries we learn that 7,321,360 cwts. of demersal fish were landed in 1905 at British ports, to which large quantity the Lowestoft trawlers contributed 276,743 cwts., valued at £302,228. The only ports at which larger quantities were landed were Grimsby, Hull, London, Milford, Fleetwood, and North Shields. Trawl-fishing is carried on all through the year.

Lowestoft also has a very important herring fishery, in which both steam and sailing drifters are engaged. During the autumn months the Lowestoft fleet is joined by a larger fleet of Scotch fishing boats, usually numbering between 500 and 600.

Herrings are caught all through the year off some part of the British coasts ; but at Lowestoft the principal season is from September till about the middle of December, when the herring shoals, in the course of their migratory movements, arrive off the Norfolk and Suffolk coasts.

In 1905 the herrings landed at the thirteen principal British herring-fishing ports weighed 3,001,167 cwts., of which 935,017 cwts. were brought into Lowestoft. The only British port where more herrings were landed was Great Yarmouth, where the catches amounted to over 1,250,000 cwts.

These figures are not easy to remember, but we will give a few more in order to show how remarkable an industry is the Lowestoft herring-fishery. At the time when this book was written, no Government report on the fisheries of 1906 had been issued, but from another source we have been able to learn the number of herrings landed at Lowestoft in that year. They numbered 25,646 "lasts." Now a last is 13,200 herrings; so we know that in one year the enormous number of 338,527,200 herrings were landed at this one Suffolk port.

In the year 1907 the Lowestoft fleets of trawlers and drifters consisted of 529 vessels, 158 of which were steamers. These vessels are distinguished by having LT painted in large white letters on their sides.

The fishing industry finds employment for a great number of men. In 1905 over 4000 fishermen were living in Lowestoft, while some hundreds more were inhabitants of neighbouring towns and villages. Besides the local men, between 3000 and 4000 Scotch fishermen had their headquarters in Lowestoft during the autumn fishing.

The packing and carting of trawl fish and herrings, together with the curing of herrings, and the preparing of them in various ways for the retail market, are also occupations in which many people are engaged. Every autumn many hundreds of Scotch girls come to Lowestoft to work in connection with the kippering of herrings, and from October to December the herring fishery is directly responsible for an addition of thousands of people to the normal population of the town.

A large proportion of the herrings caught by the Lowestoft fleet is exported to foreign countries, especially to Germany, Russia, and Italy.

The other fisheries of Suffolk are of comparatively small importance, though they find employment for many men living in the small towns and the villages along the coast.

At Southwold small boats are engaged in inshore trawling for soles and plaice, while lining for cod is also carried on. Southwold, Aldeburgh, and Thorpe fishermen catch great numbers of sprats from October to the end of January, and from April to November shrimps are caught by trawling for them. Lobsters and crabs are chiefly obtained off Thorpe and Aldeburgh.

18. Shipping and Trade.—The Chief Ports. Decayed Ports.

The principal Suffolk ports are Ipswich and Lowestoft, the former situated 12 miles from the mouth of the Orwell, and the latter on the coast.

Although Ipswich has for centuries carried on a considerable maritime trade, and in the reign of Edward III sent 12 ships to take part in the siege of Calais, it was not until about a hundred years ago that vessels of 200 tons burden could come up to its quays, and they had to be unloaded into lighters at Downham Reach, some distance below the town. To-day this port has a dock about 33 acres in extent, capable of accommodating large steamers and sailing ships. Trading steamers ply regularly between Ipswich and London, Hull, and Newcastle, while the foreign trade of the port is increasing. The principal imports are corn, oil-seed, timber, iron, and unmanufactured products for the making of artificial manures, while the chief exports are agricultural implements, railway plant, artificial manures, oils, and oil-cake.

As a fishing port Lowestoft has long been famous, but it was not until 1831 that a spacious harbour was made there by rendering Lake Lothing accessible to sea-going vessels. Subsequently a fine outer harbour was constructed, and this has since been greatly enlarged for the accommodation of the trawlers and drifters engaged in the Lowestoft

Ipswich Docks

fisheries. The South Pier is the southern boundary of the outer harbour, and adjoining it there is a dock or basin for the mooring of steam and sailing yachts during the summer months, Lowestoft being the headquarters of the Royal Norfolk and Suffolk Yacht Club.

The chief imports of Lowestoft are timber, railway plant, seed for oil-making, and coal. Formerly many Norwegian ships brought cargoes of ice to this port, but the trade in imported ice has ceased in consequence of the ice for fishing and other purposes being now manufactured in the town. During the autumn herring-fishery large quantities of herrings are exported to continental countries.

Southwold, 12 miles south-west of Lowestoft, has until lately been reckoned among the decayed ports of the Suffolk coast, but the harbour is now being dredged and enlarged for the purpose of accommodating large fishing-boats like those of Lowestoft.

Woodbridge, situated 12 miles up-stream from the mouth of the river Deben, is a small port from which corn, malt, and bricks are exported, the principal imports being coal, timber, seed and oil-cake. Only small coasting craft can make use of this port. In the eighteenth century Woodbridge was the chief port for exporting Suffolk butter.

Nearly two hundred years ago it was written of Orford that it "was once a good town, but is decayed," and although for some time after that was written Orford was a place of sufficient importance to return two members to Parliament, the heaping up of a great sand and shingle bank, which compelled the river Alde to flow some miles southward of the town before it entered the sea, resulted in the almost complete destruction of the port. It still, however, has some trade in corn and coal, and small coasting vessels ascend the Alde not only to Orford but also to Aldeburgh.

In other chapters we have referred to the almost total destruction of the once important town of Dunwich. Even as far back as the reign of Henry III its harbour had been much damaged by the sea, but the port could provide 40 ships for the use of the King. In the reign of Edward I only 11 Dunwich ships were available for naval purposes, though the port still had 16 "fair ships" and 20 barks used for trading. The harbour appears to have become quite blocked up in 1328, but butter, cheese, and corn were shipped from Dunwich to London and other ports as late as the eighteenth century.

19. History of Suffolk.

At the time of the conquest of Britain by the Romans, Suffolk, as we have seen, was inhabited by the Iceni. They appear to have submitted quietly to Roman rule for some years, but they then rose against their conquerors and were defeated. In 62 A.D. they again revolted, this time under the leadership of Queen Boadicea, who, according to a Roman historian, raised an army of 120,000 Britons, by whom three Roman towns and 70,000 Romans were destroyed.

The Saxon kingdom of East Anglia, of which Suffolk formed the southern half, was established in 575 A.D., the first king being Uffa. In the reign of Sigebert, who was killed in battle, Christianity was introduced into the county by Felix of Burgundy, and Fursey, an Irish monk who established a monastery at Burgh Castle.

During the reigns of the early East Anglian kings there were frequent conflicts between the East Anglians and the Mercians, whose kingdom bordered East Anglia on the west and north-west. In a battle fought in 635, at Bulcamp, a hamlet in the parish of Blythburgh, near Southwold, King Anna of East Anglia was slain.

In the ninth century conflicts between the rival Saxon kingdoms gave place to hard fighting against the Danes. In 865 Inguar and Ubba, with 20,000 Danes, entered East Anglia, and a battle was afterwards fought on the border of Suffolk, near Thetford, in which the Saxons were defeated. Edmund, their king, was slain, and there is a tradition that he was tied to a tree and shot to death with arrows. According to the Saxon Chronicle, Ipswich was sacked by the Danes in 991.

In the reign of Canute, Thorkill was made Governor of the East Angles. Subsequently Harold, son of Earl Godwin, was given the earldom, and he was succeeded by Alfgar, son of Leofric, Earl of Chester. He in turn was succeeded by Gurth, who was a younger son of Godwin, and who fought and died by Harold's side at Hastings.

After the Norman conquest several of the Norman followers of William I received grants of land in Suffolk. The importance of the county at this time may be judged from the fact that it occupies a considerable portion of the Domesday Book, the survey being very full and interesting.

In 1173 Robert Beaumont, Earl of Leicester, who was a supporter of the eldest son of Henry II in his plots against the king, landed an army of Flemings at Walton, near Felixstowe, and advanced into the heart of Suffolk. Haughley Castle, near Stowmarket, was besieged and taken; but at Fornham St Géneviève, near Bury St Edmunds, the invaders were defeated by the king's troops.

The Abbey Gate, Bury

About forty years later Bury St Edmunds was the scene of a great gathering of the barons and clergy of England, who, in the church of the famous abbey, vowed that they would compel King John to sign the Magna Charta.

In the thirteenth century large numbers of Flemings again arrived in Suffolk, but this was a peaceful invasion of the county, the immigrants coming to settle here and carry on the woollen industry, which flourished at Ipswich, Hadleigh, Sudbury, Lavenham, and elsewhere for several centuries.

A peasant insurrection occurred in 1381, when similar revolts took place in Kent under Wat Tyler and in Norfolk under John the Litester or dyer. The famous monastery at Bury St Edmunds was sacked, and its abbot and Sir John Cavendish, the Lord Chief Justice, were murdered.

Almost immediately after the death of Edward IV, the interest of England was for a while centred in Suffolk, for it was to Framlingham Castle that Princess Mary came to receive the support and protection of the East Anglian Roman Catholics when the Privy Council still hesitated to acknowledge her claims to the Crown. Within a few days of her arrival at the Castle, 13,000 men, ready to serve her without pay, had assembled there, while ammunition was sent by the captains of ships of war at Harwich, and at Bury St Edmunds a body of soldiers refused to fight against her. So strongly was her cause supported in Suffolk that she was soon able to enter London and ascend the throne.

During the period known as the Reformation great changes took place in Suffolk as well as in other parts of the country. By command of Henry VIII all the monasteries were closed and the monks were expelled.

Queen Elizabeth visited Suffolk on more than one

occasion, and James I was a frequent visitor, having a hunting-lodge at Thetford, where he had his headquarters while he hunted on the heaths of the north-western part of the county.

During the civil war between Charles I and the Parliament about as many of the leading men of Suffolk

Framlingham Castle

supported the one cause as the other, but no battle was fought in the county. In 1643, however, Oliver Cromwell, with five troops of horse, entered Lowestoft and made several Royalists prisoners.

In the latter part of the seventeenth century two great

naval battles were fought off the Suffolk coast. In 1665 the English fleet, commanded by the Duke of York, met the Dutch fleet under command of Admiral Opdam off Lowestoft, the latter being defeated with a loss of 18 ships taken and 14 destroyed. On May 26, 1672, the battle of Sole Bay was fought off Southwold between the allied fleets of England and France and the Dutch fleet. An old ballad writer says that the rival fleets

> "battered without let or stay
> Until the evening of that day,
> 'Twas then the Dutchmen ran away;
> The Duke had beat them tightly";

but as the losses of the English and the Dutch were about equal, victory could barely be claimed by us, although the Dutch fleet fled.

In the reign of William III a somewhat serious mutiny of troops occurred in Suffolk. When Louis XIV declared war against the Dutch, King William ordered certain regiments to be sent to the Continent to assist the Dutch army. Among these was a force of about 800 men, mainly Scotchmen, who were marched to Ipswich on their way to Harwich, where they were to embark. At Ipswich, however, they mutinied and instead of sailing for Europe they marched into Cambridgeshire. Subsequently their leaders were convicted of high treason, but the king spared their lives.

The later history of Suffolk is without any very exciting event, and chiefly concerns the agricultural and industrial development of the county.

20. Antiquities—Prehistoric, Roman and Saxon.

Antiquaries divide the prehistoric period during which Great Britain was inhabited by man into three ages. These are the Stone Age, when man had no knowledge of the use of metals and made his tools and weapons of stone; the Bronze Age, which began with the introduction of bronze-working; and the Iron Age, when man first learnt to make use of iron. At the time of the Roman conquest of Britain, its inhabitants were in the Early Iron Age. The prehistoric period ended with the arrival of the Romans.

Relics of the Stone Age are very abundant in Suffolk. In the river-gravels and brick-earths of the valleys of the Waveney, Little Ouse, Lark, and Stour flint implements of great antiquity are found. They belong to what is called the Palaeolithic or Early Stone Age, and they are usually rather large, and, in shape, either pointed or oval. Strewn over the surface of the county, and especially common on the sandy lands of the north-west and south-east, are flint implements of later date, belonging to the Neolithic or Later Stone Age. These include chipped and ground axes, delicately worked arrow-heads and spear-heads, and thousands of little tools called scrapers, most of which were used for dressing skins.

Bronze Age relics, in the form of celts (axes), swords, spear-heads and ornaments of bronze, have been found in many parts of the county. Occasionally what is called a

6—2

Palaeolithic Flint Implement
(From Kent's Cavern)

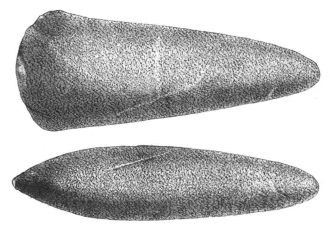

Neolithic Celt of Greenstone
(From Bridlington, Yorks.)

" bronze-worker's hoard " is discovered, usually comprising a number of bronze implements in various stages of making and sometimes accompanied by the moulds in which they were cast. Felixstowe, Martlesham, and Thorndon are among the places where such hoards have been found, while Icklingham, Mildenhall, and Lakenheath, in the north-western part of the county, have produced very many Bronze Age implements.

"Hill of Health," Culford. A Suffolk Barrow

Many of the large burial mounds called barrows or tumuli date from the Bronze Age. They are most numerous in the north-west and south-east, where much of the land is uncultivated. When these barrows are opened they are usually found to contain plain or ornamented urns which have contained the cremated remains

of the dead, but sometimes uncremated skeletons are found
in such mounds.

Only a few relics of which it can be said with certainty
that they belong to the prehistoric part of the Iron Age
have been found in Suffolk. The most important are a
few coins, a bronze-mounted tankard found at Elveden,
some horse-trappings found at Westhall, and an Italian
brooch dug up at Lakenheath. The use of iron in Britain
seems to have begun about the time of the arrival of the
Brythons, who gave their name to the country.

There are numerous ancient earthworks in Suffolk,
and some of them may date from prehistoric times. At
Bramfield, near Halesworth, there is a circular entrench-
ment in which prehistoric relics have been discovered,
while at Bungay there are slight traces of what seems
to have been an early British stockaded village. In the
neighbourhood of Icklingham some massive banks adjoining
ancient trackways are believed to be prehistoric.

The finest relic of the Roman occupation of Suffolk
is the great walled camp or fortress of *Garianonum*, at
Burgh Castle on the Waveney. The walls of this camp,
which are 14 feet high, enclose three sides of a rectangle
of 640 feet by 370 feet. There are four solid round
towers on the east side, in the middle of which is the
main entrance to the camp. This fortress was one of the
stations which were under the control of a Roman officer
called the "Count of the Saxon Shore" (*Comes littoris
Saxonici*). At Walton, near Felixstowe, there seems to
have been another walled camp, but it has been totally
destroyed by the sea.

Floors and foundations of Roman villas have been discovered at Ipswich, Icklingham, Rougham, Mildenhall, and elsewhere. Remains of Roman burials have also been met with, the most interesting being found in some artificial mounds at Rougham. On most of the Roman sites burial-urns, pottery, glass-ware and domestic utensils

Anglo-Saxon Fibulae found at Ipswich

have occurred. Great numbers of Roman coins have been discovered, and many brooches and ornaments.

Saxon relics are often found with burials of that period. Some years ago an interesting series of objects was brought to light on the site of an Anglo-Saxon cemetery at West

Stow, near Bury St Edmunds; but in 1906 a far more important find was made at Ipswich, where a Saxon cemetery of the pagan period (449 A.D. to 597 A.D.) was discovered. The objects included brooches, necklaces, drinking-cups, shield bosses, and a large number of knives and spear-heads, all of which are preserved in the Ipswich Museum. Saxon relics have also been found at Ixworth, Helmingham, and Icklingham.

21. Architecture—(a) Ecclesiastical— Churches, Abbeys and Priories.

In considering the architecture of Suffolk, it will be convenient to divide the buildings into three classes, viz.: (a) Ecclesiastical, or buildings relating to the Church; (b) Military, or Castles; and (c) Domestic, or houses and cottages. In this chapter we will deal with the Ecclesiastical buildings, commencing with the churches.

A good many of the Suffolk churches stand on sites that were occupied by churches in Saxon times, no fewer than 364 within the county limits being mentioned in the Domesday survey, but, as might be expected, very few remains of Saxon architecture are now to be seen. As the early churches fell into decay they were restored, added to, or re-built; so that now we have buildings of every period down to the present day. Modern churches, however, will not be mentioned here.

Naturally, the architecture of a county is in some degree affected by the building materials available; consequently, as flint is easily obtainable in Suffolk, it has been largely utilised, especially in beautifying the exterior of many churches with fine flint panelling or "flush-work" as it is called. Brick was used in fifteenth century building and restoration, to the detriment of the appearance of some churches. Hard chalk or "clunch" and imported

Flint panelling, Ufford Church-porch

freestone were also used. Several of the churches are thatched, and a considerable number of them have round towers. Perhaps the best tower of this kind is at Little Saxham, near Bury St Edmunds.

As we have said, very slight traces of Saxon work can now be seen in Suffolk; but at South Elmham

St Cross, near Bungay, there are ruins of a church, called the "Old Minster," which is most probably of that period. The plan of this old building is interesting because it shows that the early Christian churches were built in a similar style to the pagan temples. Traces of Saxon work

Norman Doorway, Rushmere Church

are also to be seen in the churches at Barton Mills, Gosbeck, and Debenham.

With the advent of William the Conqueror came the style of architecture we know as Norman, round-

arched, simple, massive, and imposing; at first with
but little decoration, later less massive, and with a more
liberal use of spiral and zigzag moulding. Some of the
best Norman work is in the churches at Fritton near
Lowestoft, Eyke near Woodbridge, and Polstead near
Hadleigh. The ruined choir of Orford church, too, has
some fine work of this period. Many of the churches

Early English window, Kesgrave Church

have good Norman doorways, those of Wordwell, Westhall,
and Poslingford being especially interesting. Fine Norman
work is also to be seen in the abbey gate-tower at Bury
St Edmunds. At Dunwich there is a ruined Norman
chapel of a Hospital of St James. Moyses Hall, at Bury
St Edmunds, is a Norman building which appears to
have been a Jewish synagogue.

Towards the end of the twelfth century the round arches and heavy columns of Norman work began gradually to give place to the pointed arch and lighter style of the first period of Gothic architecture which we know as Early English, especially conspicuous for its long narrow windows. It prevailed throughout the thirteenth century.

Windows of Decorated Style, Stowmarket Church

The church tower at Rumburgh, the sacristy at Mildenhall, the chancels at Exning and Hawstead, and the nave and chancel arches at Glemsford are of this period, and several other churches have good Early English portions.

The Early English led in its turn by a transitional

period into the highest development of Gothic—the Decorated period. This, in England, prevailed throughout the greater part of the fourteenth century, and was particularly characterised by its window tracery. It is well represented in this county, and especially at Woolpit, in the chancel, nave and south aisle. The nave and aisles at Orford, the chancel, the nave, and part of the tower at Dennington, and portions of the churches at Barton Mills, Ipswich (St Margaret's), Debenham, Framlingham, Stowmarket, Icklingham All Saints, and Fressingfield are good examples of this style.

The Perpendicular, which, as its name implies, is remarkable for the perpendicular arrangement of the tracery, and also for the flattened arches and the square arrangement of the mouldings over them, was the last of the Gothic styles. It developed gradually from the Decorated towards the end of the fourteenth century and was in use till about the middle of the sixteenth century. During this time some of the finest Suffolk churches were built or re-built, among them being those at Lavenham and Long Melford, which in some respects are the finest in the county. Bury St Edmunds, Ipswich, Hadleigh, Sudbury, Lowestoft, and several other towns have good Perpendicular churches, while there are many village churches built entirely or chiefly in this style. A considerable number of them have splendid roofs adorned with elaborate wood-carving ; some too, are remarkable for their carved screen-work or their fine fonts.

There were many religious houses in Suffolk before

the Reformation. They included abbeys, priories, friaries, nunneries, and hospitals, one or two of which were founded in Saxon times, while some of them were magnificent buildings, richly endowed. They were closed by Henry VIII, and allowed to fall into decay. The most famous of these monastic houses was the great

Perpendicular Porch, Ufford Church

Benedictine abbey at Bury St Edmunds. A fine Decorated gateway and a Norman gate-tower of this house are well preserved and there are other interesting portions.

There are considerable remains too, of Leiston abbey, while of Butley abbey a beautiful gateway is still standing.

At Ixworth and Clare the remains of Augustinian priories are embodied in comparatively modern houses.   Cardinal Wolsey founded a College at Ipswich, of which only a gateway remains.   Sea-wasted Dunwich possessed several monastic houses, and the ruins of a Grey Friars monastery are still to be seen there.

Leiston Abbey

22. Architecture—(*b*) **Military—Castles.**

Nearly all the ancient castles of England were built during the Norman period, consequently the original portions of these striking buildings usually have very thick walls, and the doorways semicircular arches, while such ornamentation as is preserved is characteristic of that period.   Many of them are built on large artificial mounds surrounded by a deep ditch or moat, and connected with more extensive earthworks, often covering a large area of ground.

Some were royal castles.   These were in the charge of a constable appointed by the king.   Others belonged to powerful barons, who were sometimes supporters and at other times enemies of the king.   Most of them have at some time been in the possession of the Crown, or, with the lands attaching to them, have been taken from one baron and given to another who happened to be in royal favour.

As Suffolk possesses no complete Norman castle, we will give a brief description of one.   Anyone approaching it saw first a thick and lofty wall, with towers and bastions, enclosing a considerable space of ground and surrounded by a wide and deep moat.   Spanning the moat was a drawbridge, leading to a towered gateway with a portcullis.   Through this gateway access was obtained to the outer bailey or courtyard, where were the stables.   From the outer bailey another towered gateway led to the inner bailey or quadrangle.   Within this second enclosure stood the chapel, the barracks, and the keep, which was the real fortress.   When such a castle was besieged, it was to the keep the defenders retired when the foe gained access to the inner bailey. In order that water might not fail during a siege, the keep was always provided with a well.

The only Suffolk castle retaining interesting Norman work is at Orford, where all that remains is the massive keep, which stands on a partly artificial mound surrounded by a moat. It was built about 1165, and for many years was in the charge of a constable appointed by the Crown. In 1215 it was in the possession of Hubert de Burgh—

Orford Castle

the Hubert of Shakespeare's *King John*—and from 1272 to 1276 the constable was Hugh de Dennington, who, in consequence of the depredations carried on by himself and his soldiers in the neighbourhood of the castle, has been described as a robber-baron.

D. S. 7

Framlingham Castle was one of the largest and finest in England. It was originally built by one of the Bigods, Earls of Norfolk, but it was partly demolished by Henry II in consequence of Hugh Bigod having joined with the Earl of Leicester in a revolt against the Crown. Subsequently it was held by the Uffords, Mowbrays, and Howards, from whom it passed for a while into the possession of the Crown. As we have read in our chapter on the history of Suffolk, it was at Framlingham that Princess (afterwards Queen) Mary stayed while her supporters gathered around her and took steps to place her on the throne.

The outer walls of Framlingham Castle have suffered very little during the centuries which have passed since they were built ; but all the interior buildings were pulled down many years ago. The walls are surrounded by a wide and deep moat, beyond which was an outer moat enclosing the outer bailey.

At Bungay the Bigods had another castle, of which two round towers and some massive fragments of the keep are still standing. There were Norman castles, too, at Eye, Clare, and Haughley. Of these hardly a fragment remains, but the earthworks amid which they stood are still in existence.

Mettingham Castle, near Bungay, was a fortified manor-house, built by Sir John de Norwich in the reign of Edward III. Considerable remains of it can still be seen, including a massive square gateway. Wingfield Castle, near Harleston (a Norfolk town), was another fortified manor-house, built by Michael de la Pole, Earl of Suffolk. Here, too, a gatehouse and the outer walls are the only remaining portions of the original building.

23. Architecture—(c) Domestic—Famous Seats, Manor-houses, Cottages.

Many of the finest and most picturesque houses in Suffolk were built during the Tudor period, when the comparatively settled state of the country permitted wealthy land-owners to dwell at ease on their own estates without having to occupy strongly fortified castles. During this period a number of clever foreign artists and craftsmen came and settled in England, bringing with them new ideas of building, so that by degrees the striking features of Gothic architecture disappeared, making way for classical designs in doorways and in the ornamentation of wall surfaces.

The reign of Elizabeth saw the erection of a great number of fine mansions. The ground plan of most of these was quadrangular, the main building being flanked by wings and often approached through an imposing gateway. Some houses took the shape of the letter E, perhaps as a courtly compliment to the Queen. The main entrance was usually elaborately ornamented and bore the arms of the owner of the house; while within the most striking features were the panelled hall, which had an open timber roof; the massive staircase, sometimes of stone, but more frequently of oak; the large window bays or recesses; and the minstrels' gallery. Wood and stone were largely used in the building of Tudor houses, but many were built of brick, moulded brickwork being a noteworthy feature of them.

Suffolk is rich in fine Tudor houses as well as in notable country seats of more recent date. Some of its best houses are in the neighbourhood of Bury St Edmunds,

Hengrave Hall

a town of which Daniel Defoe wrote that its beauty "consists in the number of gentry who dwell in and near it, the polite conversation among them, the affluence and plenty they live in, the sweet air they breathe in, and the pleasant country they have to go abroad in."

Hengrave Hall about four miles north-west of Bury, is a splendid house which was built about 1525 by Sir

Thomas Kytson, a wealthy London merchant. In its present state it is only about a third of its original size; but its south front and richly ornamented gatehouse are well preserved. The house is built of white brick with stone dressings.

Rushbrook Hall, about three miles south-east of Bury, is an Elizabethan house in which Queen Elizabeth was twice entertained by Sir Robert Jermyn. The room in which the Queen held her court still has the furniture it then contained. The house is moated and has the E-shaped ground-plan.

Another beautiful sixteenth-century house in the neighbourhood of Bury is Coldham Hall, in the parish of Stanningfield. It was built in 1574 by Robert Rokewode, the father of Ambrose Rokewode, who was executed for his share in the Gunpowder Plot.

At Long Melford there are two Elizabethan moated houses, Melford Hall and Kentwell Hall. At the former Queen Elizabeth was entertained in 1578 by Sir William Cordell, on which occasion, an old writer says, " He did light such a candle to the rest of the shire, that many were glad, bountifully and frankly, to follow the same example."

Gifford's Hall, in the parish of Stoke-by-Nayland, was chiefly built in the reign of Henry VIII, but some portions of it are of earlier date. Seckford Hall, near Woodbridge, was built by Sir Thomas Seckford, Master of Bequests in the reign of Elizabeth; and Helmingham Hall, the seat for many generations of the Tollemaches, a few miles from Ipswich, is a quadrangular moated house with a drawbridge which is still drawn up every night.

Country seats or manor-houses of earlier date than the Tudor period are rare, but portions of Little Wenham Hall, a brick and flint house, date from the thirteenth century and are well preserved.

Perhaps the most noteworthy and magnificent house in the county is Ickworth, the seat of the Marquess of

"Ancient House," Ipswich

Bristol. It was built between 1768 and 1803 by the fourth Earl of Bristol, who was Bishop of Derry. It consists of a large central dome, 105 feet in height and 120 feet in diameter, connected by curved corridors with two spacious wings, the length of the north front being 690 feet. The upper part of the dome is ornamented with a frieze of bas-reliefs, representing scenes from the *Iliad* and *Odyssey*.

Among the modern country houses Elveden Hall is worthy of mention. It was occupied for some years by the Maharajah Duleep Singh, who reconstructed it and decorated it in the Oriental style. Since then an Indian Hall, the details of which are modelled from ancient Indian art, has been completed.

In some of the Suffolk towns there are fine old houses, but only a few of these can be mentioned here. The "Ancient House" at Ipswich is a fine specimen of ornamental wood-work and pargetting, dating from the early part of the seventeenth century. The Moot Hall, at Aldeburgh, is a picturesque brick, flint, and timber building, known to have been in existence as long ago as the latter part of the sixteenth century; while another Moot Hall, at Sudbury, is an interesting old half-timbered house. The Shire Hall at Woodbridge dates from about 1575, and a curiously ornamented house at Bungay, probably built by some wealthy merchant, is of about the same date.

The smaller towns of Suffolk are rather remarkable for possessing quaint and picturesque houses from 200 to 400 years old.

A good deal of the charm of the rural scenery of Suffolk is due to the picturesqueness of the farm-houses and cottages, their weathered moss-grown walls and rich red tiles harmonizing well with their surroundings. Many of the old farm-houses were formerly manor-houses and a considerable number of them are moated.

A large proportion of the cottages are built of what

is called "clay-lump," the making of which is an old industry in Suffolk. After the clay is dug it is usually trodden out by a horse and mixed with spear-grass. It is then shaped into "lumps" and allowed to dry, which it does very rapidly. The cottage is erected as a framework of wood, the clay-lumps being built in, and a thatch roof is sewn on to the rafters.

Clay-lump Cottage, Barnham

Many older cottages are built of "wattle and daub." In the cases of these dwellings a rough wooden framework was first constructed and filled in with what are called "rizzes," usually consisting of hazel undergrowth, laced in and out with tar-line. This structure was then "daubed" with clay. Roofs of thatch are still put on many of the cottages. Sometimes reeds are used, but generally straw.

Very few of the modern cottages will bear comparison with the old ones in the matter of picturesqueness. The ugly little square houses with slate roofs are becoming far too common in the country districts as well as in the towns.

24. Communications—Past and Present. Roads, Railways.

The oldest roads in Suffolk are some of those which skirt the borders of the river valleys. They originated as prehistoric trackways, connecting early settlements, relics of which are often found along the sides of the valleys, where the primitive inhabitants of the country dwelt in order to be able to obtain water, fish, and wild fowl. Along the borders of the valley of the Waveney and the valleys of other rivers some of these roads are now sunk far below the level of the ground beside them, and similar hollow ways, known as "drifts," "driftways," and "lokes," lead down from them into the valleys.

Besides these valley roads, there were prehistoric highways—long trackways, usually keeping to the high ground, and generally having some connection with important settlements and camps. With one or two exceptions, these roads are not now traceable, or they have been converted into modern highways; but a well-defined ancient trackway crosses the north-western part of the county, and is known as the Icknield Way.

The Icknield Way seems to have originally connected a great prehistoric hill-fort near Dorchester with certain settlements in the county of Norfolk. It enters Suffolk at Newmarket and crosses the river Kennet at Kentford and the river Lark at Larkford, from which point it goes on across the heathland to Thetford, where it enters Norfolk.

Icknield Way, between Elveden and Lackford

The first good roads made in England were the work of the Romans, and they were constructed chiefly for military purposes. Two of their roads appear to have crossed Suffolk, and the direction taken by one of them seems to be fairly clear. It appears to have connected the Roman town of *Camulodunum* (Colchester), in Essex, with a smaller town, *Venta Icenorum* (Caistor), near Norwich, and it entered Suffolk from Essex at Stratford St Mary, on the river Stour. It then took a more or less direct course northward, leaving Ipswich on the east and Needham Market on the west, and entering Norfolk at Scole, near Diss, on the river Waveney.

Another and longer Roman road connected Colchester and Caistor, but the direction it took is a matter of dispute. The evidence now obtainable seems to prove that it lay to the eastward of the road just described, and that it connected a Roman station called *Sitomagus*, which was probably Dunwich, with Colchester and Caistor. Between Bungay and Halesworth an ancient road called "Stone Street" is believed to be a portion of this road, which as a sandy trackway can also be traced across the heathlands between Blythburgh and the little river Minsmere.

After the departure of the Romans from Britain, the fine roads they had made were not kept in repair, and for centuries after they had become almost untraversable only the roughest of roads, sandy on the heathlands and rutted or muddy on the clay lands, connected the towns and villages of the country. In 1689, however, a road suitable for coach-travelling connected London and Norfolk; for in that year Dr Rowland Davies, Dean of Cork, made the journey from London to Yarmouth by coach in two days. The road he took entered Suffolk at Newmarket and passed through Bury St Edmunds to Stuston, on the Waveney, where it crossed the river into Norfolk.

About nine years later we hear of a stage coach being

overturned in the water near Bury St Edmunds, the river, probably the Lark, being "very high, a great flood." In 1724 roads of that part of the county were infested by highwaymen; for in the Diary of William Coe, of Mildenhall, quoted by Dr Raven in his *History of Suffolk*, we are told that in that year the sister and the daughter of the diarist were "persued" near Wangford "by a foot padd & were forced to gallop almost to Eriswell to escape."

Early in the eighteenth century turnpike roads were made between the principal towns of Suffolk, and by 1734, when a survey of Suffolk was made by John Kirby, the author of the *Suffolk Traveller*, the county was crossed by several highways, connected with many cross-roads and byways.

At the present time the County and District Councils are responsible for keeping the roads in repair, and since they have had charge of them they have been greatly improved.

Of the two principal roads from London, one crosses the river Stour at Stratford St Mary, about 10 miles south of Ipswich, while the other crosses the western boundary of the county at Newmarket and branches to Thetford and Bury St Edmunds.

The Great Eastern is the principal railway. The main line from London enters Suffolk about 10 miles south of Ipswich and crosses the eastern part of the county to Lowestoft, sending off branch lines to Felixstowe, Aldeburgh, Southwold, Hadleigh and Framlingham. At Beccles the main line is connected by the Waveney

Valley branch with another line which passes through
Mid Suffolk, and which in turn is connected at Haughley
Junction with lines to Bury St Edmunds, Newmarket
and several towns in the south-western part of the county.
The Midland and Great Northern Joint Railway, in
conjunction with the Great Eastern, has a short coast line
between Lowestoft and Yarmouth.

There is no canal of any importance in Suffolk, but
the rivers Stour, Orwell, Deben, and Waveney are
navigable for several miles to small coasting craft, barges,
and wherries. Passenger steamers ply daily on the Orwell
between Ipswich, Harwich, and Felixstowe, while small
river steamers make pleasure trips on the Waveney during
the Broadland season.

25. Administration and Divisions— Ancient and Modern.

The government of England is partly central and
partly local. The Parliament sitting in London makes
laws for the whole country, but it has given counties
and portions of counties the power to manage many of
their own affairs. This system of government is, in the
main, a development of that which was organised in
Saxon times. Then the king was the supreme ruler, and
there was also a kind of central parliament, consisting of
bishops, abbots, and the principal owners of land. By
these authorities the whole country was governed, but
then, as now, there were local authorities to deal with

local matters. The principal local authority was the shire-mote, which was responsible for the government of the shire or county. It met twice a year, after Michaelmas and Easter, and it consisted of the freeholders of the county. Its chief officers were the Ealdorman and the Sheriff, the latter representing the king, and being present to guard the rights of the Crown.

The counties were divided by the Saxons into "Hundreds," each of which was originally supposed to consist of a hundred families of freemen. The county of Suffolk was divided into 22 hundreds. Each hundred had its own court, the Hundred Mote, which met once a month, for the purpose of discussing the affairs of its own district and judging any disputes which had arisen between the inhabitants of the hundred. Some central place was usually chosen for the holding of its meetings. Sometimes it met on the side of a hill called the mote hill or beneath some well-known tree. Some of the Suffolk hundreds, like Lackford and Wangford, took their names from river fords ; others from small meres or lakes.

There were yet other local authorities in Saxon times. Each hundred consisted of a number of townships, or, as we should call them, parishes, and each of these townships had its own court or *gemot*, at which any freeman could appear. This court met whenever necessary under the presidency of an officer called the reeve, and it was its business to make laws for the government of the township and to see that the laws were obeyed. At Bungay an officer called the town reeve is still elected every year as chairman of a body having control of the town lands.

Suffolk was also divided into what are called "Geldable portions" and "Franchises." Geldable portions are those over which the king holds the chief rights, while the Franchises are those in which the chief rights are held by the lords.

The Suffolk Franchises are the Franchise or Liberty of St Etheldred, now belonging to the Dean and Chapter of Ely; the Franchise of St Edmund, which belonged to the abbey of St Edmundsbury; and the Liberty of the Dukedom of Norfolk.

If we now describe the present mode of government in Suffolk, we shall see how it resembles and how it differs from the earliest mode of administration.

In the first place, the county has two chief officers known as the Lord Lieutenant and the High Sheriff. The former is generally a nobleman or rich landowner, who is appointed by the Crown, while the latter is chosen every year on November 12.

The central form of county government is the County Council. In Suffolk there are two County Councils, one for East Suffolk and the other for West Suffolk, the former holding its meetings at Ipswich and the latter at Bury St Edmunds. These councils consist of aldermen and councillors, whose business it is to keep the main roads and bridges in repair; deal with matters of finance, allotments and small holdings; manage asylums; and carry out other important work. Special committees are appointed for different purposes. Large committees, including several co-opted members, deal with matters relating to education.

County Councils were established by Act of Parliament in 1888. In 1894 another Act was passed by which the conduct of much of the business of large parishes was placed in the hands of Urban and District Councils, while the smaller parishes were given Parish Councils. In Suffolk there are 10 Urban and 18 District Councils.

Ipswich Town Hall

The towns of Ipswich, Bury St Edmunds, Lowestoft, Aldeburgh, Beccles, Eye, Southwold, and Sudbury are municipal boroughs having the power of managing their own affairs. Ipswich is also a county borough.

Suffolk is also divided into Poor Law Unions, each of which is under a Board of Guardians, whose duty it is to manage the workhouses and appoint officers to carry on the work of relieving the poor.

For purposes of justice the county has two Courts of Quarter Sessions and it is divided into 19 Petty Sessional Divisions, each having magistrates or justices of the peace, whose duty it is to try cases and punish offenders against the law. Ipswich, Bury St Edmunds, Lowestoft, Aldeburgh, Eye, Southwold, and Sudbury have their own magistrates, while Ipswich, Bury St Edmunds, and Sudbury have separate Courts of Quarter Sessions.

There are 519 civil parishes in Suffolk, but for ecclesiastical purposes the county is divided into 465 parishes. Of the ecclesiastical parishes, 303 are within the Diocese of Norwich, and 162 are in that of Ely. For the better ecclesiastical government of the county, these parishes are divided into two archdeaconries and 22 rural deaneries.

Suffolk is represented in the House of Commons by eight Members of Parliament. It is divided into seven parliamentary constituencies, five of these being county divisions, each represented by one member. Ipswich returns two members and Bury St Edmunds one member.

26. The Roll of Honour of the County.

Suffolk has been the native county or place of residence of a great number of famous men, and it stands foremost among the counties in respect to the production of men of marked distinction and intellectual ability. If we look through the pages of the *Dictionary of National Biography*, we find that for several centuries Suffolk has given to England a considerable proportion of her greatest men. It is impossible even to mention the names of all these

Cardinal Wolsey

Suffolk worthies in this chapter, but we will make brief reference to some of the most prominent of them.

Many famous divines have been born in Suffolk. Bishop Grosseteste, famous for his opposition to the Papal usurpation, was born at Stradbroke, near Eye; Archbishop

Sancroft, who was ejected from Lambeth Palace for refusing to take the oath of allegiance to King William III and Queen Mary, was born and is buried at Fressingfield; while Archbishop Tybald, who was beheaded by Wat

George Crabbe

Tyler's insurrectionists, was a native of Sudbury. Richard Sibbes, a famous Puritan divine, born in 1577, was the son of a Tostock wheelwright. The great Cardinal Wolsey was born in 1471 at Ipswich, where his father

8—2

was a tradesman, following, among other avocations, that
of a butcher.

Among the Suffolk poets George Crabbe takes, perhaps,
the foremost place.   He was born at Aldeburgh in 1754,
and his poems contain faithful descriptions of the scenery

Edward FitzGerald

in and around his native town.   Henry Howard, the poet
Earl of Surrey, was born at Tendring Hall, in the parish
of Stoke-by-Nayland.   He was beheaded in the reign of
Henry VIII, and is buried in Framlingham church.   John

Lydgate, the poet-monk of St Edmundsbury and a contemporary of Chaucer, took his name from his native Suffolk village, Lidgate. Honington, near Thetford, was the birthplace, in 1766, of the peasant-poet Robert Bloomfield, who as a lad worked on a farm in the neighbouring parish of Sapiston. Thomas Nashe, a gifted but dissolute sixteenth-century satirist, was born at Lowestoft.

Among later poets and men of letters Edward Fitz-Gerald, who was born in 1809 at Bredfield, near Woodbridge, and who died in 1883, is famous for his delightful letters to his friends and for his translation of the *Rubáiyát* of Omar Khayyam.

Suffolk has been the native county of several famous sailors and admirals. Noteworthy among them are Thomas Candish or Cavendish, of Trimley near Ipswich, who was the second Englishman to sail round the world; and Sir Philip Broke, who was born at Nacton, and who, while in command of the *Shannon*, fought and captured the American frigate *Chesapeake* within sight of Boston Harbour in 1813. Another Suffolk man who distinguished himself beyond the seas was John Winthrop, the founder of the State of Massachusetts. He was born at Groton, near Hadleigh, in 1588.

Among famous lawyers, three Lord Chief Justices, Sir John de Cavendish, Sir Robert Wright, and Sir John Holt, were members of Suffolk families.

Two of England's greatest artists were natives of this county. Gainsborough, the landscape and portrait painter, was born at Sudbury, and John Constable, whose landscapes are among the chief art treasures in our National

Gallery, was the son of an East Bergholt miller. In Thomas Woolner, who was born at Hadleigh in 1826, Suffolk can claim a poet and sculptor of some renown.

John Constable

Several women who distinguished themselves in literature are entitled to a place in the roll of honour of this county, but we must be content with naming Agnes Strickland, the historian of the queens of England, who was born at Reydon, near Southwold, and Mrs Inchbald, the actress and novelist, born near Bury in 1753.

A considerable number of famous men have been intimately connected with Suffolk, though they were not born in the county.

Charles Dickens visited Ipswich, Bury St Edmunds, and Lowestoft, and some of the most amusing incidents described in the *Pickwick Papers* are supposed to have happened in the two first-named towns, while Blundeston, near Lowestoft, was chosen to be the birthplace of the hero of his favourite novel, *David Copperfield*.

George Borrow, the author of the *Bible in Spain*, *Lavengro*, and other well-known books, spent many years of his life at Oulton, where he died in 1881. The house in which he lived has been pulled down, but the summer-house in which some of his books were written is still standing beside Oulton Broad.

Bury St Edmunds is rich in literary associations. Daniel Defoe, the author of *Robinson Crusoe*, lived there for some time; Samuel Taylor Coleridge, the poet, was a visitor there, and Oliver Goldsmith was occasionally a guest at Barton Hall, near Bury, the home of his intimate friends the Bunburys. The Bury Grammar School is famous for having numbered among its scholars a great many boys who afterwards distinguished themselves in various ways, and several of whom were natives of Suffolk. Among them were Sir Thomas Hanmer, the first editor of Shakespeare ; J. M. Kemble, the historian of the Anglo-Saxons; James Spedding, the editor of Bacon ; and Frederick Malkin, the historian of Greece.

William Kirby, the entomologist, was born at Witnesham Hall, and educated at Ipswich; while last, but by no means of least importance on our list must come Arthur Young, " the first to raise agriculture to a science," the travelled observer and voluminous writer on his subject, who, though not born in Suffolk, lived most of his life there, at Bradfield, where he lies buried.

27. THE CHIEF TOWNS AND VILLAGES OF SUFFOLK.

(The figures in brackets after each name give the population in 1901, and those at the end of each section are references to the pages in the text.)

Aldeburgh (2405) is a popular watering-place lying between the sea and the estuary of the river Alde. From 1571 to 1832 it returned two members to Parliament, but it decayed as a coast town and port in consequence of the encroachment of the sea. It was the birthplace of George Crabbe, the poet. (pp. 37, 38, 44, 74, 76, 103, 112, 113, 116.)

Beccles (6898) is pleasantly situated on the river Waveney. Although an inland town, it was formerly a fishing port, herrings being brought up the river and landed there. Its chief industries are printing and the manufacture of agricultural and marine engines. (pp. 67, 112.)

Blythburgh (646), on the river Blyth, was formerly a town and port of some importance, but it fell into decay owing to the silting-up of the river. Its church is one of the finest in the county. (p. 78.)

Brandon (2327), on the Little Ouse, is chiefly noted for its ancient flint-knapping industry, which is described in the chapter on the industries of the county. (pp. 56, 67.)

Bungay (3314) is an ancient and picturesque town on the Waveney. The ruins of a Norman castle, built by one of the Bigods, Earls of Norfolk, stand in the centre of the town, adjoining some remains of extensive earthworks. A Benedictine nunnery was founded here in 1160 by Roger Glanvile and his wife. The town possesses a quaint old market cross, surmounted by a figure of Justice. There are large printing works here. (pp. 67, 86, 98, 103, 110.)

Burgh Castle (527), a village situated at the point where the river Waveney enters Breydon, the estuary of the Norfolk river Yare, is the *Garianonum* of the Romans. There are massive remains of a large Roman camp here. (pp. 77, 86.)

Bury St Edmunds (16,255), the principal town in West Suffolk and the third in size in the county, stands on rather high ground sloping down to the bank of the river Lark. It owes its name to King Edmund of East Anglia, who was killed by the Danes. His body was interred in the church of a monastery founded by King Sigebert in the seventh century, and this monastery subsequently became one of the wealthiest and most famous abbeys in England. Great numbers of pilgrims came to visit the shrine of St Edmund, among them being several English kings. Parliaments and councils of clergy were held in the abbey, and it was in the abbey church that the English barons took an oath that they would compel King John to sign Magna Charta. The principal remains of the great abbey of St Edmundsbury are a lofty Norman tower, a fine Decorated gateway, portions of the abbey church, and a picturesque thirteenth-century bridge known as the Abbot's Bridge. Bury also possesses two of the finest churches in Suffolk. In St Mary's church, Mary Tudor, Queen of France and afterwards wife of Charles Brandon, Duke of Suffolk, is buried. She died in 1533, and was first interred in the abbey church. Moyses Hall, which now contains a Museum, is a well-preserved Norman building, which was originally a Jewish synagogue. Portions of the Guildhall date from the fifteenth century. Bury is the depot of the Suffolk Regiment. The chief industry of the town is the making of agricultural implements and malting and brewing machines. (pp. 54, 63, 79, 80, 91, 93, 94, 100, 111, 112, 113, 119.)

Butley (270) is a village about three miles west of Orford. Its interesting feature is a fine gateway of an Augustinian priory, founded in 1171. This gateway now forms a part of the vicarage. Michael de la Pole, Earl of Suffolk, who was killed at Agincourt, was buried in the priory church.   (p. 94.)

Carlton Colville (2375) is a small town, portions of which adjoin Lowestoft.   Part of the parish borders Oulton Broad.

Clare (1582), the ancient seat of the Earls of Clare, is an interesting little town on the north bank of the Stour.   Here are extensive ancient earthworks, amid which stood a Norman castle, slight traces of which can still be seen.   There was also an Austin priory here, in the church of which Joan of Acres, second daughter of Edward I and Queen Eleanor, was buried.   She was born at Acres while her father was fighting the Saracens, and she married Gilbert Earl of Clare.   (pp. 95, 98.)

Corton (618) is a coast village about two miles north of Lowestoft.   A good many visitors resort to it in the summer. (pp. 42, 48.)

Debenham (1182) is a small town near the source of the Deben.   Its church is one of the few in Suffolk believed to contain Saxon work.   (pp. 90, 93.)

Dunwich (157), a few miles north of Southwold, is now a small fishing village.   Formerly it was a large and important town and port; but it has been almost entirely destroyed by the sea.   (pp. 38, 39, 45, 46, 77, 91, 95, 107.)

East Bergholt (1397), a charming village in the valley of the Stour, was the birthplace of the artist John Constable, whose father lived at Flatford Mill.   (pp. 11, 118.)

Eye (2002) is a municipal borough in the northern part of Suffolk.   A Norman castle was built here by a member of the Malet family; it stood on a huge mound, which, with some other earthworks, can still be seen.   (pp. 98, 112, 113.)

Felixstowe (3507) is a popular watering-place about two miles north of Harwich harbour. Together with Walton it forms an urban district. As a health resort this town is in favour on account of its sheltered position. (pp. 35, 36, 44, 52, 54, 55, 69, 79, 85, 86.)

Fornham St Géneviève (92) a village about three miles north-east of Bury St Edmunds, is the scene of the defeat of the army of Flemings, which, under command of the Earl of Leicester, landed on the Suffolk coast in 1173. (p. 79.)

Framlingham (2526) is famous for its castle, the outer walls of which are so well preserved that it hardly has the appearance of a ruined stronghold until the wrecked interior is seen. This castle was built by the Bigods, Earls of Norfolk. For many years it was in the possession of the Howards, Dukes of Norfolk. Framlingham church contains some magnificent tombs to members of this family, among them being that of Henry Howard, Earl of Surrey, who was beheaded in the reign of Henry VIII. (pp. 80, 93, 98, 116.)

Fritton (270) is a village near the border of a lake which is one of the most picturesque sheets of water in the Broads district. (pp. 19, 32, 91.)

Glemsford (1975) is a large village north of the Stour. Many of its inhabitants are engaged in silk-weaving and coconut mat making.

Gorleston (13,710) is partly situated in the administrative county of Norfolk. It is a popular and rising coast resort at the entrance to Yarmouth harbour. It is partly built at the north end of a line of cliffs extending from Lowestoft, and it has a fine sandy beach. (p. 42.)

Hadleigh (3245), an ancient market town on the little river Brett, was formerly an important centre of the Suffolk cloth-weaving industry. In 1635 it had 47 tradesmen engaged in various branches of the cloth trade against 47 of all other trades combined. Some of its houses are old and picturesque, dating from the fourteenth, fifteenth and sixteenth centuries. Near the church is a fine fifteenth-century gatehouse 43 feet high. Dr Rowland Taylor, rector of Hadleigh, and for a time chaplain to Archbishop Cranmer, was burnt at the stake on Aldham Common, near this town, in the reign of Queen Mary. (pp. 65, 93, 118.)

Halesworth (2246) is a market town on a small tributary of the river Blyth.

Haughley (789) is a junction station on the Ipswich, Bury and Norwich lines. There are some ancient earthworks in the village, including a large mound on which a Norman castle was built. In the reign of Henry II this castle was surrendered to the Earl of Leicester, who landed an army of Flemings on the Suffolk coast. (pp. 79, 98.)

Havergate Island (4) is a tract of land near the entrance to Orford Haven. It is about two miles long and surrounded by the river Alde or Ore. It is used as a sheep farm.

Haverhill (4862), a town situated in the extreme south-west of the county, is noted for its manufacture of textile fabrics. About half of its population is engaged in making horse-hair cloths, rugs, mats and clothing.

Hoxne (838) is a pretty village on the south bank of the Waveney. Traditionally it is the scene of the martyrdom of King Edmund of East Anglia, who is said to have been tied to a tree and shot to death with arrows by the Danes.

Icklingham (339), a township of two parishes on the river Lark, is interesting to antiquaries on account of the great number of Early British, Roman, and Saxon antiquities which have been found there. Prehistoric flint implements are very abundant in and around the village, and remains of Roman villas have been discovered. An ancient road or trackway, the Icknield Way, crosses the heathlands of this place. (pp. 85, 86, 87, 88, 93.)

Ipswich (57,433), the chief town of Suffolk, is a county and parliamentary borough, situated on the northern slope of the valley of the Gipping and Orwell, the latter name being given to the tidal part of the river. It is a port of considerable trade. The discovery of floors of Roman villas proves that there was a Roman settlement here, while the fact that coins were minted here in Saxon times suggests that it was then a place of some importance. According to the Saxon Chronicle, the town was plundered in 991 by the Danes, and at the time of the Norman Conquest it had nine churches. It was granted its first charter in 1199. Evelyn, who visited it in 1677, wrote of it that "it has in it fourteen or fifteen beautiful churches; in a word, it is for building, cleanness, and good order, one of the best towns in England." Cardinal Wolsey was born here, and founded a college for Secular Canons; its gateway is still in existence. One of the most interesting buildings in the town is the "Ancient House," a picturesque and richly ornamented house built in the early part of the seventeenth century. At the present time Ipswich has 17 churches, some of which have Norman portions. The Town Hall is a fine building with a clock tower. It contains portraits of several of the kings and queens of England. Other noteworthy buildings are the Corn Exchange, the Museum, the Free Library, the Shire Hall, and the Custom House. Christchurch Park, in the midst of which stands an Elizabethan house, consists of about 50 acres of wooded park land and is now a public park. (pp. 5, 63, 65, 74, 78, 82, 87, 88, 93, 95, 103, 111, 112, 113, 115, 119.)

Ixworth (856) is a village on the bank of a tributary of the Little Ouse. An Augustinian priory was founded here about 1100. Some portions of it are embodied in Ixworth Abbey, a modern house. (pp. 88, 95.)

Kersey (482), a village in the valley of the Brett, gave its name to the fabric known as kerseymere. (p. 65.)

Kessingland (1420) is a large fishing village about four miles south of Lowestoft.   It has some popularity as a watering-place.   (p. 41.)

Kirkley (6465) is the southern portion of Lowestoft.   (See **Lowestoft.**)

Lavenham Church

Lavenham (2018) is a small town standing on high ground near the Brett, a tributary of the Stour.   It was formerly a centre of the cloth-weaving industry.   Its church is one of the finest parish churches in England.   There are several old half-timbered houses in the town.   (pp. 65, 93.)

Leiston (3259) is a growing town about four miles north-west of Aldeburgh.   Large iron-works are its principal industry. There are very picturesque ruins of Leiston abbey, a Premon-stratensian monastery, founded in 1182 by Ranulph de Glanvile, who took a prominent part in the battle with the Scots at Alnwick, and who accompanied Richard I to the Holy Land. (pp. 63, 94.)

Long Melford (3080) is an interesting old town on the banks of a small tributary of the Stour. In the fifteenth century it was an important centre of the woollen industry. Its late Perpendicular church is a very fine one. Melford Hall and Kentford Hall, in this town, are well-preserved Elizabethan houses. (pp. 93, 101.)

Lowestoft (29,850), the second largest town in Suffolk, and one of the most popular watering-places in England, is the easternmost town in the kingdom. It is a municipal borough, and one of the chief British fishing ports. It has a long and spacious sea-front, two fine piers, a large harbour, and several large docks or basins for the accommodation of its big fleets of fishing vessels. At the north and south ends of the town there are wide parades for visitors; there are also two small but very pleasing public parks. The older portion of the town is built on the summit of a line of cliffs from which the sea receded some centuries ago; but the more modern portions are chiefly on lower ground near the harbour and Lake Lothing and at the north and south ends of the borough. The parish of Kirkley, on the south side of the harbour, is included in the borough, and it is the part of the town chiefly resorted to by visitors. Extensive fishmarkets adjoin the harbour on its north side. The principal industries of the town are its fisheries, net-making, the building of fishing-vessels, herring-curing, and the building of motor-cars. (pp. 4, 5, 42, 47, 48, 49, 51, 52, 54, 55, 65, 71, 72, 73, 75, 76, 81, 82, 93, 112, 113, 117, 119.)

Mettingham (318), a village near Bungay, is noteworthy on account of its ruins of a castle or fortified manor-house dating from the reign of Edward III. (p. 98.)

Mildenhall (3567) is a town and very large parish on the border of Cambridgeshire. About one half of the parish was once a

Herring Market, Lowestoft

Fenland mere and a considerable portion of it is sandy heathland. Mildenhall seems to have been the site of large settlements of Stone Age races, flint implements of various ages having been found here in great numbers. Relics of the Bronze Age and the Roman and Saxon periods have also been found. (pp. 85, 87, 92, 108.)

Newmarket (10,686). A portion of this town is in the county of Suffolk, in which county the whole town is included for administrative purposes. It is a great horse-racing centre and of late years it has grown considerably in consequence of its popularity with sporting men. (p. 8.)

Orford (885) is a decayed coast-town on the west bank of the river Ore or Alde. It possesses a massive keep of a Norman castle built in the reign of Henry II. (pp. 19, 36, 37, 44, 50, 51, 76, 91, 93, 97.)

Oulton (1860), near Lowestoft, is a village and large parish extending down to the shore of Oulton Broad. (pp. 19, 119.)

Pakefield (1425), a coast village adjoining the borough of Lowestoft on the south, has suffered considerably of late years in consequence of sea encroachment. Its cliffs are continually falling and many houses have had to be pulled down. (pp. 46, 47.)

Saxmundham (1452) is a small market town on the main line from London to Lowestoft.

Shotley (750) is a scattered village at the point of the peninsula between the estuaries of the Stour and the Orwell. A fourteen gun battery here commands the entrance to Harwich harbour.

Southwold (2800) is a municipal borough and favourite seaside resort at the mouth of the river Blyth. It stands on fairly high ground and is connected by a ferry with the picturesque village of Walberswick. The town was incorporated by Henry VII. (pp. 39, 40, 46, 48, 51, 74, 76, 82, 112, 113.)

Stowmarket (4162) is a market town at the junction of three rivulets uniting to form the Gipping. Its old vicarage was the residence of Dr Thomas Young, the tutor of the poet Milton. An ancient mulberry tree in the garden of this house is said to have been planted by the poet. Gun-cotton, cordite and other explosives are manufactured in the town. (pp. 65, 93.)

Sudbury (7109), on the river Stour, is an ancient town which formed part of the estate which Earl Morcar forfeited to William the Conqueror. Camden, the sixteenth-century historian, writes that "Sudbury, the South Burgh, men suppose to have been the chief town of the shire, and to have taken its name in regard of Norwich, which is the northern town. Neither would it take it well at this day to be counted much inferior to the towns adjoining, for it is populous and wealthy by reason of clothing here." This reference is to the woollen industry established by Flemish immigrants in the reign of Edward III. For many years Sudbury was famous for its manufacture of bunting and baize, but this industry has given place to the making of silken goods and coconut matting. The town has two interesting churches and several picturesque old houses. It was the native place of the artist Gainsborough. (pp. 65, 66, 93, 103, 112, 113, 115.)

Walberswick (304) is a very picturesque coast village separated from Southwold by the river Blyth. (p. 39.)

Wickham Market (1417) is a small town with a station on the main line from London to Lowestoft.

Wingfield (387), a village five miles south of Harleston, in Norfolk, is noteworthy on account of its remains of Wingfield Castle, a fortified manor-house dating from the fourteenth century, when it was built by Michael de la Pole, Earl of Suffolk. (p. 98.)

Woodbridge (4640) is an old town delightfully situated on the river Deben. Formerly it was a port of considerable trade, sending ships laden with corn to Scotland and Ireland. Daniel Defoe, who visited the town in 1722, states that it was the chief port for the "shipping off" of Suffolk butter. Barges and other small coasting craft still carry cargoes between this town and other ports. The Shire Hall is a quaint old building dating from 1575, and there are some old half-timbered houses. An Augustinian priory was founded here about 1193, but no recognizable trace of it remains. The poet Crabbe was for a time apprenticed to a Woodbridge doctor. (pp. 76, 103.)

Wrentham (1019) is a small town on the road from Lowestoft to Southwold.

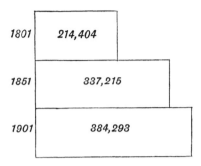

Fig. 1. Diagram showing increase of population of Suffolk

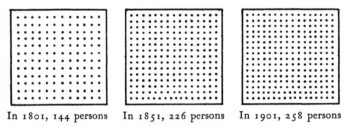

In 1801, 144 persons   In 1851, 226 persons   In 1901, 258 persons

Fig. 2. Population of Suffolk to the square mile

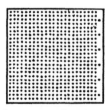

In 1901, 558 persons

Fig. 3. Population of England and Wales to the square mile

Fig. 4.　Diagram showing comparative areas of Suffolk
and the rest of England and Wales

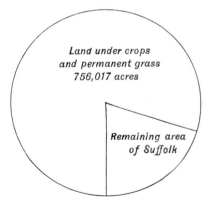

Fig. 5.　Proportionate area of cultivated and uncultivated
land

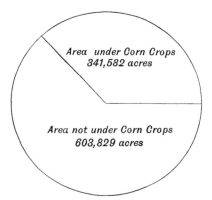

Fig. 6. Area of Corn Crops compared with that under
other cultivations in Suffolk

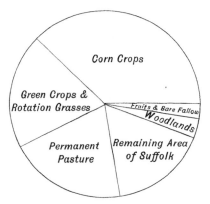

Fig. 7. Proportionate areas of cereals, pasture, crops,
woodlands, etc.

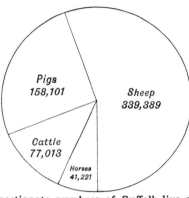

Fig. 8.   Proportionate numbers of Suffolk live stock in 1907

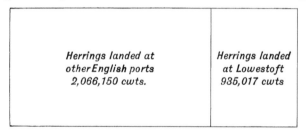

Fig. 9.   The Lowestoft herring catch compared with that
of the remaining English ports

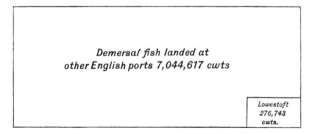

Fig. 10.   The Lowestoft catch of demersal fish compared
with that of all other English ports

www.ingramcontent.com/pod-product-compliance
Ingram Content Group UK Ltd.
Pitfield, Milton Keynes, MK11 3LW, UK
UKHW042146280225
455719UK00001B/136